Renewable Energy Transition with Artificial Intelligence

Challenge-driven Solutions

Nina Dethlefs and Joyjit Chatterjee

Loughborough University School of Science
Department of Computer Science
Epinal Way
UK

WILEY

Published by John Wiley & Sons, Inc., Hoboken, New Jersey.

Library of Congress Cataloging-in-Publication Data has been applied for:

Hardback ISBN: 9781394300037
ePDF ISBN: 9781394300051
epub ISBN: 9781394300044
oBook: 9781394300068

Cover Design: Wiley
Cover Image: © Fei Yang/Getty Images

Contents

List of Figures

List of Tables

Preface

The high-voltage distribution and transmission of electricity is associated with a range of environmental effects that are increasingly contentious and lead to difficult land-use planning issues. In the United Kingdom, local planning authorities have been taking an anticipatory approach to such difficulties through the formation of policy addressing these issues in their statutory plans. This is generally with the aim of setting constraints on the location and form of high-voltage installations and nearby development.

This book describes an in-depth research project into this emerging body of policy in local authority development plans. It explores the specific concerns being raised with regard to high-voltage installations and the influences that shape policy formation. This is approached by means of a two-level investigation: firstly, a study of policy content throughout England and Wales and, secondly, an in-depth examination of policy formation in a number of case study authorities across England and Wales.

The research reveals a number of consistently occurring policy concerns and themes with regard to high-voltage installations that focus explicitly on the protection of local environmental conditions. This is seen as an indication of broader trends apparent throughout plan-making in the last decade, but is also understood to result from the interaction of a wide range of localised features and the close involvement of certain actors. It is, however, a highly contested policy area in some respects and is therefore subject to considerable modification through plan-making procedures. This suggests that there is scope for a more cooperative manner of addressing the issues raised by the presence and planning of high-voltage installations and of establishing relevant policy. A number of recommendations are made that aim to facilitate a greater understanding of different perspectives by the parties most involved in this policy area and to improve the planning of high-voltage installations and related development.

Acknowledgements

I would like to thank Professor Christopher Wood of the School of Environment and Development at the University of Manchester for his thorough supervision of the research described here and to Mr Hector Pearson and Mrs Catherine McCloskey of National Grid Transco for their encouraging and helpful support. I would also like to thank the many people who gave considerable time and assistance to the work carried out for this research, especially the staff of Malcolm Judd and Partners, and the planning officers, other local authority staff, elected members and voluntary group members throughout the United Kingdom.

This book is based upon a research project established jointly between the University of Manchester and National Grid. It was undertaken as an ESRC Collaborative Studentship Award, number S00429937039.

Abbreviations used in the text

AONB	Area of Outstanding National Beauty
AVDC	Aylesbury Vale District Council
BCBC	Bridgend County Borough Council
CIGRE	International conference on large high-voltage electric systems
CP	'Countryside protection'
CPRE	Council for the Protection of Rural England
DETR	Department of the Environment, Transport and the Regions
DNO	Distribution network operator
DoE	Department of the Environment
DPP	Development plan process
DTI	Department of Trade and Industry
DTLR	Department of Transport, Local Government and the Regions
EC	European Community
EIA	Environmental impact assessment
EiP	Examination in public
EMF	Electromagnetic fields
ER	'Encouraging regeneration'
FoE	Friends of the Earth
HVDT	High-voltage distribution and transmission
kV	Kilovolt
LDDC	London Docklands Development Corporation
LP	Local plan
LPA	Local planning authority
LPI	Local public inquiry
LTBC	Langbaurgh on Tees Borough Council

MDC	Mendip District Council
NC	Newham Council
NCC	Norfolk County Council
NG	National Grid
NRPB	National Radiological Protection Board
OHL	Overhead line
PCB	Polychlorinated biphenyl
PPG	Planning Policy Guidance
RAP	'Residential area protection'
RCBC	Redcar & Cleveland Borough Council
RCEP	Royal Commission on Environmental Pollution
REVOLT	Rural England Versus Overhead Line Transmission
RMBC	Rotherham Metropolitan Borough Council
SBC	Swindon Borough Council
SNPA	Snowdonia National Park Authority
SoS	Secretary of State
SP	Structure plan
TBC	Thamesdown Borough Council
TC	Tynedale Council
UDP	Unitary development plan
UK	United Kingdom

CHAPTER 1

Introduction

1.1 THE DEVELOPMENT OF HIGH-VOLTAGE SYSTEMS

Electricity has become a fundamental form of energy in industrialised societies. Its use is steadily increasing, not only in those countries that are in earlier stages of developing energy-dependent systems, but also in those countries whose energy consumption is already high (International Energy Agency 2001, World Bank 2000).

The use of electricity depends on the transfer of electricity from a point of generation to a point of consumption. Although it is possible for generation to take place on a localised level for local use, integrated systems of electricity generation and transfer have now developed on regional, national and even international scales. This has been brought about by the increasingly centralised organisation of the electricity industry, in which nationalisation has played a major part (Patterson 1999). Large power stations are now generally connected to long-distance transmission networks, which then feed into more localised networks of electricity distribution for direct supply to consumers. These interconnected systems of transmission and distribution are generally judged to be more economic and reliable than the unconnected local systems that they replaced (Messing *et al.* 1979, National Grid 1999a).

Transmission and distribution systems consist mostly of linear features: overhead power lines supported on towers or poles of various kinds, and underground insulated cables used mostly for lower voltage distribution. Substations are also located at points where voltage changes occur. The longer-distance transfer of electricity is achieved most efficiently at high voltage; this requires the equipment involved to be of a relatively large scale. As systems have extended to cover larger areas, the voltages used also have increased; typically, a transmission system operates on two or three voltage levels between 100 and 500 kV (kilovolts), sometimes higher.

High Voltage Electricity Installations S. Jay

The transmission component of an electricity network is generally referred to as a *grid*. In the United Kingdom, a national grid system was first established in the 1930s, operating at 132 kV. This was upgraded in the 1950s, with the introduction of a "supergrid" operating at 275 kV, which was, in turn, upgraded in the 1960s to a system incorporating 400 kV equipment. The remaining components of the original 132 kV system subsequently became part of the system of electricity distribution to consumers, leaving the 275 kV and 400 kV equipment as the "transmission" system proper (Carruthers 1987). This supergrid was developed under public ownership of the electricity industry, by the Central Electricity Generating Board.

In 1990, the UK electricity industry was privatised and broken up into its component parts. The transmission grid in England and Wales became the responsibility of National Grid (NG), while the distribution systems were handed over to 12 regional electricity companies (now referred to as *distribution network operators* (DNOs)) (different arrangements were applied in Scotland and Northern Ireland). In England and Wales, therefore, NG now owns all 275 kV and 400 kV equipment (Figure 1.1), and the DNOs own nearly all 132 kV equipment, along with lower voltage equipment for electricity distribution to customers. The transmission network currently allows excess electricity generation in the north and midlands to be exported to the south of England (National Grid 1999a).

This book will concentrate on, for reasons given below (Section 1.5), the land-use planning concerns of local authorities regarding:

- the 275 kV and 400 kV transmission grid and
- the 132 kV element of the distribution systems

throughout England and Wales. For convenience, the combined components of transmission and high-voltage distribution will be referred to throughout this text as "high-voltage distribution and transmission" (HVDT) (see also Section 1.5). This term will usually refer to the elements of the electricity network throughout England and Wales specified above. It is also occasionally used (in the following section, for example) to refer to the equivalent elements of electricity networks more generally throughout the world, although precise voltages vary between different systems.

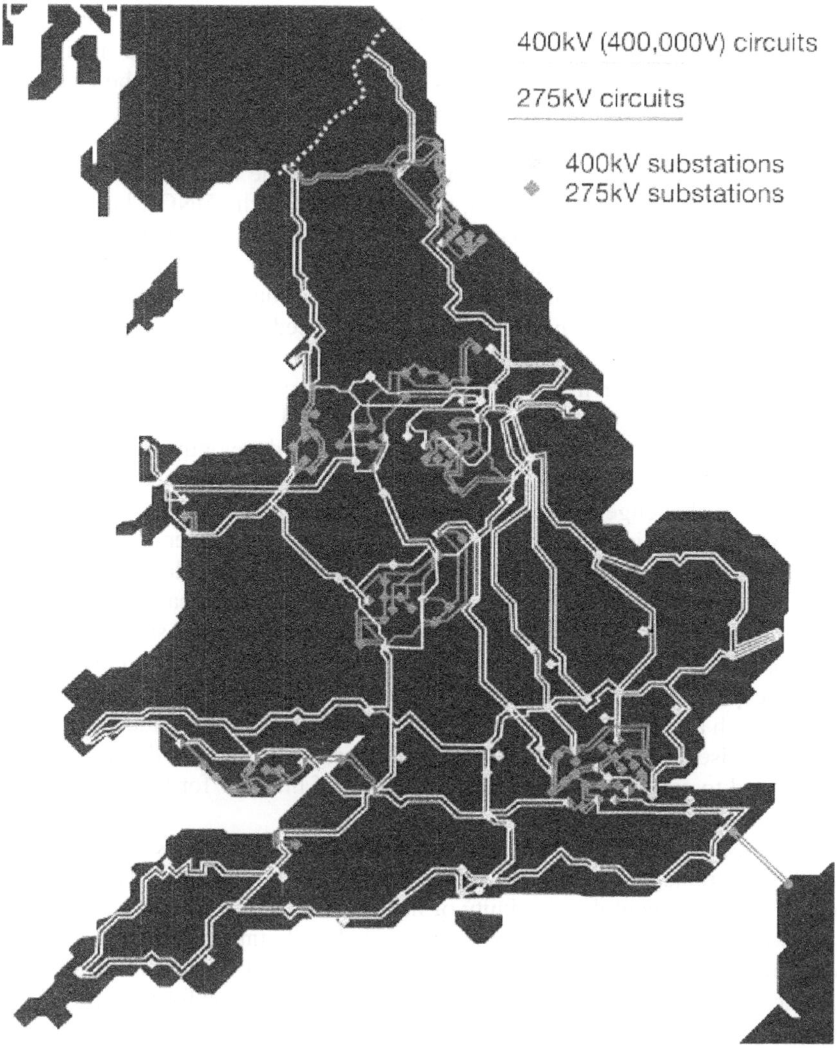

Figure 1.1 The present transmission system in England and Wales (Courtesy National Grid Company unpublished.d)

1.2 LAND-USE PLANNING RELATING TO HVDT INSTALLATIONS

Given their extensive nature and scale, the siting of HVDT installations is a major land-use planning exercise. Planning procedures are usually necessary both in relation to the construction of new equipment and significant replacement or upgrading of existing equipment. These procedures generally involve a range of bodies. Authorisation may lie at the level of local, state, regional or national government; procedures for gaining consent may include consultation with various organisations and the participation of affected communities and individuals; agreements may need to be reached with landowners regarding the placing of equipment and access for both its construction and maintenance.

The development of HVDT systems has generally been justified on the basis of public service provision. In the United Kingdom, the instrumental bodies, the Central Electricity Generating Board, and now NG and the DNOs, have operated as statutory undertakers; when seeking planning consents, they have been able to call upon the notion of "public utility", the provision of essential services (Ernst 1994), in support of their applications. The proposed routeing and location of new or replacement installations has been largely influenced by technical and economic matters, but has also had regard to environmental considerations. However, over the last decade or two, in many parts of the world, increasing attention to environmental issues is being paid during the planning of HVDT equipment. This reflects growing disquiet with the adverse effects of industrial developments generally, but also indicates increasing public concern regarding HVDT in particular. This trend has acted as a challenge to the assumption that "public utility" can be used as an overriding argument in favour of proposed developments (Rogier *et al.* 1996).

The particular public concerns that have been raised in relation to electricity HVDT are focused on two main areas. Firstly, the direct physical effects of installations, the visual effects of overhead lines and their supporting structures, and of substations, have been of most concern in this regard, particularly in areas of valued landscape and in residential areas. Other related matters have also been given attention, such as ecological, archaeological, recreational, effects and so on. These have given rise to opposition to new installations, particularly in designated areas, and have led to calls for the alternative routeing of lines, or for them to be placed underground.

The second main area of concern has been the perceived risk to human health from the electromagnetic fields (EMF) generated by power lines. This has led to growing anxieties about the proximity of lines to other forms of development, especially where people spend significant periods of time, such as residential development. These anxieties have, like the visual and other effects mentioned above, led to opposition towards proposed HVDT installations; for example, in areas where they would be placed near to housing. But they have also provoked opposition to proposed other development, such as housing, where it would be in close proximity to *existing* HVDT installations. This issue therefore opens up questions regarding the planning not just of HVDT equipment but also of other forms of development that must take into account the presence of HVDT installations.

Concentration on these issues has led to increasing public attention being paid to planning applications relating to HVDT. This has typically led to protests from local communities, often represented by interest groups and local politicians, and has involved local authorities in disputes with proponents of certain projects. With regard to proposals for projects involving new installations, for example, this has led to increasing lengths of time being required for the planning phase. For instance, a new transmission line project in the Netherlands requiring only two years for construction involved a 14-year lead time for planning, owing to local opposition and consequent problems in obtaining consents (Adema *et al.* 1996). Electricity utilities throughout the world are increasingly reporting difficulties of this kind, and are, as a result, turning their attention to technical means of alleviating some of the environmental effects of their proposed actions (e.g. Bascom *et al.* 1996, Bohme *et al.* 1998, Doyen-Viguier *et al.* 1996).

1.3 LOCAL PLANNING AUTHORITIES AND HVDT INSTALLATIONS

In the United Kingdom, local planning authorities (LPAs) have a key role to play in the land-use planning issues outlined above. Applications regarding overhead lines (above a certain voltage) are determined by central government (currently by the Secretary of State for Trade and Industry), under the Electricity Act 1989. The Act provides, however, for LPAs to act as statutory consultees with respect to applications for overhead lines in their areas. Moreover, should an LPA raise any objection to a proposed line, the Secretary of State is obliged to call a public inquiry into the application. LPAs have, therefore, a significant, if not determining, part to play in the planning of high-voltage overhead lines. Proposals for new electricity substations, on the other hand, come before LPAs as planning applications under the Town and Country Planning Act 1990; LPAs are therefore authorised to approve these installations. Similarly, the planning of other forms of development that may be affected by the presence of existing electricity installations, such as housing or schools, normally comes within the domain of ordinary planning applications, for which the LPA may grant approval.

One of the means by which LPAs in the United Kingdom have recently been seeking to bring their influence to bear on planning matters relating to HVDT has been the introduction of specific policy in this regard. This is seen as an anticipatory measure providing LPAs with their own guidance on any future applications concerning HVDT that may affect them. This has been in the form of policies and supporting statements incorporated into statutory development plans that the local authorities are required to prepare as a framework for land-use planning decisions in their areas Department of the Environment, Transport and the Regions (DETR) 1999a). The last decade or so has therefore seen an increasing number of local authorities establishing, or attempting to establish, policy relating to HVDT through the preparation of their development plans.

The use of development plans as an expression of LPA attitudes on this particular set of issues has coincided with the renewed emphasis that has been placed more generally on development plans in the planning system over this period. The Planning and Compensation Act 1991 asserted the importance of development plans in local planning decisions, and set off a new wave of plan preparation in England and Wales (Adams 1994). This instrument has therefore provided a means by which LPA concerns regarding HVDT installations have been articulated.

The policies and supporting statements relating to HVDT that have been drawn up during the preparation of development plans have concentrated on the concerns outlined above (Section 1.2). They have sought primarily to protect local environmental quality rather than, for example, to make a contribution to the establishment of an efficient national electricity network. As such, proposed policies and supporting statements have become the focus of some conflict with other interests, especially sections of the electricity industry and, occasionally, the housebuilding industry and other developers. Struggles over development plan wording have taken place in a number of the processes of plan preparation, through representations and responses made during consultations, and also through evidence presented at public inquiries and examinations in public. In general, groups opposing LPAs, especially representatives of the electricity industry, have been successful in bringing about changes to proposed policies and supporting statements, making them more acceptable to their interests.

1.4 THE FORMATION OF HVDT-RELATED POLICY BY LPAS

There is, therefore, a growing trend for LPAs in the United Kingdom to advance policy through their development plans that aim to defend the local environment against the effects of HVDT installations. The emergence of this body of policy provokes a number of questions about the particular concerns being raised and the influences at work in the formulation of policy. For example, what are the issues that are being addressed by LPAs, and why? To what extent do they reflect broader environmental concerns, or are they principally related to local circumstances? Can policy formation be explained in terms of geographical, social, economic, and other factors? Who are the principal actors involved in policy formation? Does group or individual activism play a major part? Why is it that certain LPAs advance this policy area while others do not? This book explores these questions, with the overall aim of investigating the processes by which LPA policy on planning aspects of HVDT has recently been coming to expression. This book therefore presents a study on the formation of local policy on a discrete and environmentally sensitive subject, while at the same time raising a number of highly topical issues relevant to wider environmental planning.

In general terms, public policy can be thought of as a political statement, or "a set of political purposes" (Parsons 1995, p. 14), the formation of which is a complex process, subject to many possible influences and the involvement of many different actors. Consequently, a host of theoretical understandings has arisen of the ways in which public policy comes to expression (Hill, 1997). Adherence to a particular model of policy-making is not, however, a prerequisite for the empirical study of a specific policy area; indeed, presupposing the means by which policy is formed may constrain an empirical study too tightly. Nonetheless, some kind of conceptual framework of policy formation is necessary in order to organise the investigation of interacting elements (John 1998).

With regard to understanding LPA policy that deals with the effects of HVDT installations, particular attention has been given to

- the external features being brought to bear upon the process of policy formation, and their inter-relation;
- the role of actors, whether institutions, groups or individuals, and the interaction between them;
- the concerns expressed in specific policy content, and their inter-relation;
- the formal procedures by which concerns are brought to expression in development plans.

These elements are represented diagrammatically in Figure 1.2. This figure suggests that various geographical, social, environmental and other features that are external to the immediate policy process are influential upon actors involved in formulating policy. Their concerns and interests are given expression through the content of more specific policies and statements, which form part of development plan documentation. This overall process is both initiated by and mediated through the formal procedures of preparing development plans.

This book therefore describes research into the emerging formation of HVDT-related policy as advanced by LPAs through the preparation of their development plans. It investigates in detail the motivating interests and forces that have brought about and shaped this developing body of policy throughout England and Wales. In addition, the essentially empirical work described here gives rise to a generalised model of policy formation, which has the potential for wider applicability. The research was carried out between 2000 and 2002, and was based on the expression of policy in development plans being prepared, or that had recently been completed, at the time of the analysis.

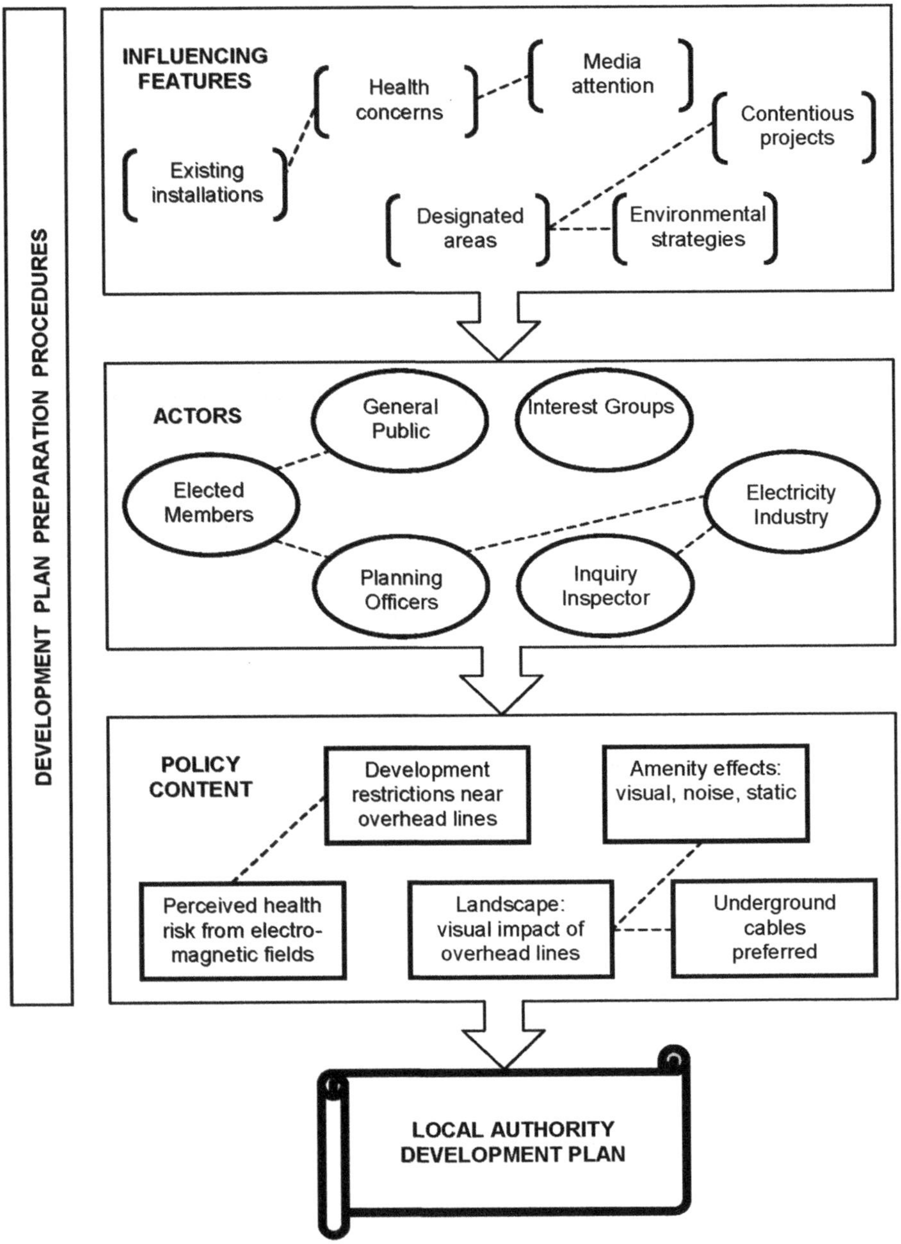

Figure 1.2 A representation of the formation of HVDT-related policy

1.5 SCOPE AND ORGANISATION OF THE BOOK

The geographical scope of the research described here was determined firstly by the need to consider as wide a range of LPAs as possible, both in the sense of policy positions adopted by LPAs, and (given the variety of forms of LPA following local government reorganisations) in the sense of different types of planning authority. A national, rather than regional, level of study was therefore chosen, in order to ensure a wide coverage of possible influences on policy formation. The second aspect of the geographical scope of this research is the nature of the electricity industry. As mentioned above (Section 1.1), the transmission grid in England and Wales became a separate enterprise with privatisation of the industry in 1990, NG becoming the relevant licence holder. This transmission system (along with its associated supply and distribution systems) provides a convenient boundary for this study, incorporating the nine regions of England and Wales. Working within NG's area of responsibility also provides an institutional focus to the study, which is itself relevant to policy formation, as LPAs in England and Wales will often have NG specifically in mind (along with more local components of the industry) when drawing up policy.

Local authority policy is expressed in statutory development plans. A development plan is a document that sets out a local authority's policies and proposals for the use of land in their area (DETR 1999a); different types of plan are prepared by different authorities. This research is based upon HVDT installations that have been treated as a policy area in the following types of plan in England and Wales:

- structure plans, prepared by county councils and some unitary and national park authorities in England;
- unitary development plans, prepared by London boroughs, metropolitan and some other unitary authorities in England, and most Welsh unitary authorities;
- local plans, prepared by district councils, national park authorities and some unitary authorities in England.

The system of plans is, in fact, currently undergoing radical reform (DTLR 2001a), but as this has yet to be fully implemented, this research focuses on policy formation within these existing types of plan (waste and minerals local plans were not considered).

In addition to containing policies and proposals, a development plan document also contains the local authority's reasoning for its policies. This is referred to as "*reasoned justification*" in the case of unitary development and local plans, and "*explanatory text*" in the case of structure plans. These two forms of supporting statements do not carry the same status, as government guidance indicates that the explanatory text of a structure plan does not form part of the plan (though it may be included in the document); this is in contrast to the reasoned justification of unitary development and local plans (DETR 1999a). Despite this lack of equivalent legal standing to reasoned justification, explanatory text does carry an equivalent function of enlarging upon the wording of policies.

More generally, both a development plan's "policies" on a particular subject and the accompanying supporting statements are expressive of a local authority's policy on that matter. It is, therefore, the statements as a whole relating to HVDT within development plans that are the focus of this book.

As indicated above (Section 1.1), the topic areas of this book are both electricity transmission and the highest voltage level of distribution, referred to as *HVDT*. This is because local authority policy tends to take as its area of concern installations operating at 132 kV and above; this suggests that 132 kV equipment is perceived to be the threshold at which problematic planning issues are encountered. This does not equate with the electricity industry's demarcation between transmission and distribution, but this is of no great significance to land-use planning issues as viewed by LPAs. Furthermore, wording in some development plans suggests that LPAs are not always aware of the distinction drawn by the electricity industry between "transmission" and "distribution". The policy scope of this research is therefore taken to be policy relating to HVDT installations as defined above.

The following chapters are organised as follows. Chapters 2 and 3 explore two distinct contexts within which HVDT-related policy is situated. Chapter 2 considers HVDT as an industrial sector. It begins by looking at the legislative procedures governing the land-use planning of HVDT installations, with emphasis on those provisions designed to ensure environmental protection. This leads on to an examination of the environmental effects that are associated with HVDT, some of which underlie the concerns currently being expressed by LPAs. Chapter 3 turns to the means by which these concerns are being articulated, the system of development plan preparation. The current importance of development plans is investigated, as are the trends and priorities that have been emerging in the practice of plan-making. The importance of plan preparation as a process, with considerable opportunity for the involvement of different actors, is underlined. These two chapters therefore provide an understanding, firstly, of the policy topic itself, and, secondly, of the procedures by which it is being advanced.

Chapter 4 sets out a methodological approach for the investigation of HVDT-related policy throughout England and Wales, on the basis of the framework shown in Figure 1.2. A two-level approach is developed, consisting of a study of policy content across the whole of England and Wales, and an in depth examination of policy formation in a number of localities. These are seen as complementary means of building up an overall picture of LPA concerns and the influences that are shaping them.

Chapters 5 to 9 describe the empirical work that is at the heart of the research described in this book. Chapter 5 sets out the results of the first level of study, identifying patterns in HVDT-related policy across England and Wales as a whole. This provides the basis for the second, more qualitative, phase of study, the examination of policy formation in 10 case study authorities. These are individually described in Chapters 6 to 8. Chapter 9 presents an analysis of the body of case studies, and identifies a number of key policy themes in common between them.

Chapter 10 draws on the results of the empirical work as a whole, and discusses the dominant elements and influences found to be at work in the formation of HVDT-related policy. This discussion refers back to the contexts described in Chapters 2 and 3, and also to wider literature and policy developments relevant to the particular points in question. Finally, Chapter 11 considers the overall forces that are acting upon the formation of HVDT-related policy. Recommendations are also made by which LPA concerns might be better accommodated.

CHAPTER 2

High-voltage Distribution and Transmission in England and Wales

2.1 INTRODUCTION

Analysis of a particular policy area requires knowledge not just of the process of policy formation, but also of the sector to which it relates (John 1998). An understanding is needed of both the legislative and administrative framework within which the sector operates, and those technical aspects of the sector that policy may touch upon. This chapter therefore presents an understanding of the context of, and issues relevant to, the policy area in question.

High-voltage distribution and transmission (HVDT)-related policy, as advanced by local planning authorities (LPAs), is primarily concerned with the environmental implications of HVDT installations within LPAs' areas (1.4). The background to this topic is therefore provided by:

- the provisions that apply to the land-use planning of HVDT installations; and
- the environmental effects of HVDT activity.

The first half of this chapter sets out a summary and discussion of the legislation and guidance governing the consent procedures for HVDT installations. This not only provides knowledge of the relevant procedures, but also indicates that the nature of the procedures may play a role in the shaping of LPA policy on HVDT.

The second half of this chapter explores the environmental issues raised by HVDT activity. This involves a review of documents produced by the electricity industry covering the environmental aspects of their operations. This presents a particular perspective on the environmental effects of HVDT activity, with which the concerns raised by LPAs can be compared.

High Voltage Electricity Installations S. Jay

THE LAND-USE PLANNING OF HVDT INSTALLATIONS

2.2 CONSENT PROCEDURES FOR HVDT INSTALLATIONS

HVDT installations are of three kinds (1.1), which are dealt with under different consent procedures:

- overhead lines (OHLs) require consent from the Secretary of State, under the Electricity Act 1989;
- substations require planning permission from LPAs, under the Town and Country Planning Act 1990;
- underground cables are classified as 'permitted development' under the Town and Country Planning (General Permitted Development) Order 1995, and do not require individual consent.

These provisions are outlined in greater detail and commented upon below. Requirements for the environmental assessment of certain HVDT projects are also described. Finally, safety standards affecting the planning of HVDT installations are mentioned, as is a draft circular on the planning of developments giving rise to electromagnetic fields (EMF).

HVDT-related policy refers not just to the location of HVDT installations themselves, but also to the siting of other forms of development, such as housing, in proximity to existing installations (1.3). These developments normally come within the usual planning system, and are dealt with directly by LPAs under the Town and Country Planning Act 1990.

2.3 THE ELECTRICITY ACT 1989

2.3.1 The Duties of Licence Holders

The Electricity Act 1989 (referred to as 'the Act' in the remainder of this chapter) provided for the privatisation of the electricity industry, and covers consent procedures for electricity generation and transfer. Under the Act, the Secretary of State (currently for Trade and Industry) may grant licences for the distribution (actually referred to as 'supply' in the Act) and transmission of electricity (s.6). In England and Wales, a transmission licence is currently held by the National Grid (NG), and distribution licences are held by the distribution network operators.

The Act states the general duties of licence holders. These are as follows:

- the development and maintenance of "an efficient, coordinated and economical system" of electricity distribution and transmission;
- "to facilitate competition in the supply and generation of electricity" (s.9).

The first of these refers to the function of providing an integrated system of electricity transfer from generators to customers. It is, however, qualified by the requirement to do so efficiently and economically. The second duty, requiring licence holders to facilitate competition in the industry, is understood to mean that they must connect any new generating plant to the grid system (NG 2001a).

Both of these duties have significant implications for the planning of new HVDT installations. Firstly, licence holders must give considerable priority to the economics of installation provision. This will encourage them to choose more direct line routes, and also to avoid costly methods of connection, such as placing lines underground. Secondly, licence holders must connect any new power plants to the existing grid system regardless of the location of those new schemes, the distances to be covered by new lines, or the nature of the land that will be affected. In other words, the duties upon licence holders contain obligations that have inevitable, possibly contentious, land-use consequences.

However, the Act also places certain environmental responsibilities on licence holders with implications for the siting of new installations. These are covered by "Preservation of Amenity and Fisheries" (s.38 and Schedule 9) and are described below (2.3.3). There is, therefore, some degree of balance within the Act between the general duties discussed above and the environmental aspects of the electricity industry's activities; the extent to which one prevails over the other with regard to any given proposal has the potential for considerable disagreement between different parties.

2.3.2 Section 37 Consent

The consent mechanism for new OHLs (except for some of low voltage, and those exempted under certain regulations) is described in Section37 and Schedule 8 of the Act, and is known as Section 37 consent. Key steps of the procedure are as follows:

- an application is made to the Secretary of State (SoS);
- 'relevant' LPAs are notified (those through whose areas the OHL would pass);
- LPAs may object to the application to the SoS;
- if an objection is made by an LPA, the SoS must call a public inquiry;
- the SoS may also call a public inquiry in the light of other objections or material considerations;
- the SoS makes a decision regarding the application.

The only means for any party to appeal against the SoS's decision is to call for a judicial review through the courts; however, this action is limited to claiming that the SoS has operated outside the bounds of the law.

When granting Section 37 consent, the SoS may also "direct that planning permission for that development ... shall be deemed to be granted" (Schedule 8, para.7(1)). This is also allowed for under the Town & Country Planning Act (s.90). This 'deemed planning permission' is, in effect, the simultaneous conferring of both consent and planning permission by the SoS.

These procedures are also described in Circular 14/90, which "gives advice about the procedure for obtaining the views of local planning authorities in cases where the consent of the Secretary of State for Energy, together with deemed planning permission, is sought ... to install or keep installed an electric line above ground" (DoE 1990a, para. 1).

The consent mechanism for OHLs therefore lies primarily in the hands of the Secretary of State, with only a limited role for LPAs. This is broadly in line with planning procedures for public utilities generally, where it can be argued that the obligation to provide services to the public requires a stronger role to be played by the central government (Cullingworth and Nadin 2002). The fact that these special procedures have been maintained despite the privatisation of the electricity industry suggests that electricity provision continues to be seen primarily as a public utility rather than just a form of private, entrepreneurial development. The decisive role played by the Secretary of State may also be indicative of the linear, long-distance nature of OHLs; as they frequently cross over several (sometimes many) local authority areas, central, rather than local, control could be seen as necessary to ensure their effective development.

The separation of OHL consent procedures from the planning system is to some extent balanced by the effective power of relevant LPAs to bring about a public inquiry into any application. This ability, plus the participation that LPAs would then have in any public inquiries, gives them a significant voice in proposals for OHLs in their areas. Moreover, it is likely that the electricity industry would prefer to enter into discussion with affected LPAs rather than face the delays and difficulties that would be caused by public inquiries. LPAs are therefore in a strong position to negotiate conditions making proposed lines more acceptable to them, even if they cannot determine the overall application.

2.3.3 The Amenity Clause

The Act includes amongst the duties of licence holders the 'Preservation of Amenity and Fisheries', in what is known as the Amenity Clause (Section 38 and Schedule 9). This requires that a licence holder:

(a) shall have regard to the desirability of preserving natural beauty, of conserving flora, fauna and geological or physiographical features of special interest and of protecting sites, buildings and objects of architectural, historic or archaeological interest; and

(b) shall do what he reasonably can to mitigate any effect which the proposals would have on the natural beauty of the countryside or on any such flora, fauna, features, listed building or objects (schedule 9 para.1).

An earlier version of the Amenity Clause first appeared in legislation of the 1940s, as a safeguard against the possible damaging effects of major hydroelectric power schemes in Scotland and Wales. It was then adapted for inclusion in the Electricity Act of 1957 (Section37), specifically to offer protection against the "spidery trail of pylons" (Sheail 1992, p. 158) that was affecting the countryside with the development of the supergrid (Section 1.1). This placed a duty upon the nationalised electricity industry of the time to pay regard to the preservation of amenity in words very similar to those quoted above; it remained the only statutory responsibility of its kind to be placed upon a utility for some years (Carruthers 1987).

The use of the word 'amenity' to describe the clause is a reflection of the conservationist interests being expressed at the time of the clause's formation; the Council for the Protection of Rural England (CPRE) was particularly active in promoting its inclusion in legislation (Sheail 1992). The requirement to preserve flora, fauna, landscape, buildings, and so on for their 'beauty' and 'interest' can be seen as typifying attitudes towards the environment before the growth of modern environmentalism in the 1970s (Newby 1990); for instance, post-war attitudes towards the countryside have been described as thinking of it as an aesthetic backcloth (Healey and Shaw 1993).

The drafting of a new Electricity Bill, in preparation for privatisation of the industry, provided an opportunity to reconsider the Amenity Clause. However, little change was made to its 1957 wording; this was despite lobbying from the CPRE, the Countryside Commission and others, who argued that the Act should now address wider environmental issues. The only significant change to the clause that they succeeded in bringing about was that, whereas the clause had originally said that 'account should be taken' of any effect on the countryside, and so on, licence holders were now required to 'mitigate' any effect (Sheail 1992).

The reference to mitigation, although no more than a minor rewording of the clause, is nonetheless an introduction into the Act of a more recent 'environmental management' perspective to the environment (Healey and Shaw 1993). It is a term frequently used in the context of more comprehensive approaches to environmental protection (such as environmental impact assessment (Section 2.5)), and does represent a more rigorous approach to the possible effects of proposals. However, the clause remains restricted to the preservation of natural features and human artefacts, with no interest expressed in other environmental protection issues, such as pollution and human health. This is despite the serious polluting effects of certain components of the electricity industry, especially generation, which are also covered by the Act. Overall, the Amenity Clause now appears to be antiquated and overly limited in its scope.

The Amenity Clause is followed in the Act by a requirement that each licence holder produce "a statement setting out the manner in which he proposes to perform his duty" under the clause (para. 2(1)). This is known as a 'Schedule 9 statement'. It is also stipulated that the statement should be prepared (and periodically reviewed) in consultation with a number of bodies with statutory conservation duties; currently these are the Countryside Agency, English Nature, the Countryside Council for Wales, English Heritage and CADW (Welsh Historic Monuments).

2.4 OTHER PROVISIONS RELATING TO CONSENT

2.4.1 The Planning System: Substations

The land-use planning of substations does not fall within the remit of the Electricity Act, but is covered instead by the conventional planning system administered by LPAs under the Town and Country Planning Act 1990. This may seem inconsistent, given that substations are an integral part of grid systems, and are characterised by the convergence of OHLs from different directions; moreover, they are often large-scale installations creating similar planning issues to OHLs. However, substations are unlike OHLs in being 'point' rather than linear features, and are therefore more localised in nature. This makes proposals for new substations more suited to the planning system. However, the existence of different consent regimes for integrated units of electricity systems does create the potential for conflicting decisions being reached, and raises the question of whether a more strategic approach to planning would be preferable.

2.4.2 Permitted Development: Underground Cables

Certain electricity operations are classified as 'permitted development' and do not require any applications for consent or planning permission. Permission is automatically granted under the Town and Country Planning (General Permitted Development) Order 1995 (Part 17, class G); here, the electricity utilities operate as statutory undertakers. The Order applies mainly to relatively minor operations, such as the replacement of lines; however, it also covers the installation of all underground cables. This may be understandable in relation to low voltage cables, particularly those supplying individual consumers. But high-voltage cables are also included, despite their being major installations with significant environmental effects (Table 2.3); they are more equivalent to oil and gas pipelines (for which consent is required) than to low voltage distribution lines. The lack of any consent procedures for their installation therefore appears to be an anomaly. In practice, however, relatively few high-voltage lines are placed underground; when they are, this is often following consultation with LPAs.

2.5 ENVIRONMENTAL IMPACT ASSESSMENT

Under the European Union legislation (EC Directives 85/337, 97/11), proposals for new OHLs are often subject to environmental impact assessment (EIA). Lines with a voltage of 220 kV or above and a length of at least 15 km compulsorily require EIA (Annex I para. 20). Shorter or lower voltage lines may be subject to EIA according to their potential environmental impacts (Annex II para. 3b).

In England and Wales, these requirements are enforced through regulations (Electricity Works (Environmental Impact Assessment) (England and Wales) Regulations 2000). These define 'Schedule 1' projects as OHLs where EIA is compulsory (as specified by the Directive), and 'Schedule 2' projects as those where EIA may be required, if the proposed line:

- has a voltage of 132 kV or more, or
- is to be installed in an area designated for its landscape, conservation, and so on interest.

Judgements regarding the need for EIA in these cases depend not just on these criteria, however, but are made on a case-by-case basis, taking into account such things as the length of the line, visual impact and associated development (Department of Trade and Industry 2000). The ultimate decision regarding the need for EIA in Schedule 2 cases (the 'screening decision') rests with the Secretary of State; relevant LPAs also have an opportunity to put forward their views. For all OHLs where EIA is required, an environmental statement must be submitted along with the application for Section 37 consent.

EIA has been defined as "the systematic identification and evaluation of the potential impacts of proposed projects ... relative to the physical-chemical, biological, cultural and socioeconomic components of the total environment" (Canter 1996, p. 2). It involves a comprehensive assessment of all the likely environmental effects of a development. This typically uses the techniques of environmental science (O'Riordan 1995), and also places emphasis on consultation with expert bodies and participation by the affected public (Wood 2003). As such, EIA can be seen as a tool of environmental management, which adopts an analytical approach to a proposal and seeks to guide the decision-making process accordingly (Barrow 1999).

In relation to the planning of HVDT installations, EIA requirements introduce a new perspective. Unlike the Electricity Act, they demand a comprehensive and rigorous analysis of the environmental effects of a proposal, and incorporate formal consultation and participation procedures. They represent the introduction of an environmental management approach to the planning of OHLs, which can be seen as a considerable development from the relatively narrow 'natural beauty' and preservationist concerns of the Amenity Clause.

It should be noted that EIA has provided a model that has been used for HVDT proposals where it is not a statutory requirement. NG, for example, has undertaken to provide environmental statements with applications for all major overhead lines; this is mentioned in the government circular referred to above (DoE 1990a, para. 6). NG also carries out EIA-type assessments for many other projects such as substations and underground cables.

2.6 SAFETY STANDARDS AND DRAFT EMF CIRCULAR

2.6.1 Safety Standards

Given the hazardous nature of high-voltage equipment, certain safety regulations and guidance deal with the location of HVDT installations and nearby development. These are concerned with maintaining minimum distances between electrical equipment and people, to prevent potentially fatal 'flashover' occurring. For example, the Electricity Supply Regulations 1988 specify the minimum heights at which overhead conductors should be strung. The electricity industry has also established its own minimum clearances between conductors and nearby buildings, trees, roads, and so on (Electricity Association 1988). Clearly, these are relevant not only to the planning of HVDT installations, but also to that of subsequent development nearby (NG 2001b).

2.6.2 Draft EMF Circular

One of the principal environmental issues arising from the location of HVDT installations is the possible effect of exposure to EMF upon human health (1.2; Figure 2.1). In response to public concern about the alleged health consequences of both OHLs and telecommunications equipment, the government produced a draft circular in 1998 which aimed to guide LPAs on the matter (Department of the Environment, Transport and the Regions and Department of Health 1998). This stated the following:

- the government's statutory advisor on EMF issues is the National Radiological Protection Board(NRPB);
- the NRPB does not accept a proven causal link between EMF exposure and the incidence of cancer;
- operators are under legal responsibilities to keep EMF levels below certain levels;
- LPAs proposing policies that would introduce 'cordons sanitaires' around OHLs, and so on should have regard to the above points;
- public perception of danger is a valid planning consideration, although the weight attached to this will depend upon the case in question.

However, this draft did not proceed beyond the initial consultation stage. It is likely that this is due to the ongoing nature of research into the possible health effects of EMF since 1998; the findings of important research projects were awaited. Major reports have now been published (including into the possible effects of EMFs emitted by mobile phone base stations) (NRPB 2001, Independent Expert Group on Mobile Phones 2000), confirming that there is no strong evidence of a link between EMF and cancer. However, these reports do call for further research to be conducted in certain areas, reflecting a continuing uncertainty regarding the possibility of health risk, particularly for children, resulting from prolonged EMF exposure.

THE ENVIRONMENTAL EFFECTS OF HVDT INSTALLATIONS

2.7 PERSPECTIVES ON THE ENVIRONMENTAL EFFECTS OF HVDT

The environmental effects of HVDT installations feature prominently in HVDT-related policy. An understanding of the range of effects associated with this industrial sector therefore provides a context within which LPA concerns can be placed; this section presents such a context. However, the environmental effects of HVDT cannot be described as an indisputable set of actions upon the environment; they are defined and understood in different ways by different groups, and given varying degrees of importance. This is illustrated by the changing focus in legislation, as discussed above, from 'amenity' considerations to a wider environmental management approach. Any description of effects will therefore be an expression of certain preconceptions about their nature and significance.

Nonetheless, it is possible to provide a broad picture of the environmental effects associated with HVDT, acknowledging that it will be shaped by certain interests. Here, one particular perspective on the effects of HVDT is given, that of the electricity industry. This perspective is chosen for two reasons: firstly, the industry provides the most comprehensive description of effects available, through a body of commercial and technical literature; secondly, it will be seen that the industry's views play a significant role in the formation of LPA policy.

The electricity industry's understanding of the environmental effects of HVDT is presented here by a review of UK and international documentation:

- corporate environmental reports, produced by UK electricity companies;
- environmental statements, produced for UK HVDT projects;
- guidelines for the planning of HVDT installations, produced by UK electricity companies;
- papers presented to CIGRE (International Conference on Large High Voltage Electric Systems), a worldwide, industry-based technical organisation.

2.8 CORPORATE ENVIRONMENTAL REPORTS

An increasingly common corporate practice is the production of annual environmental reports. These summarise companies' stated performance in environmental aspects of their operations, often measured against specific targets, and also set goals for future performance. This is in line with current 'environmental management' practice (Hunt and Johnson 1995, Netherwood 1996). Most of the UK electricity utilities now produce these reports, which reveal what they consider to be the significant environmental issues relating to their sector.

Table 2.1 provides a summary of the effects dealt with in electricity utility environmental reports produced by NG, Northern Electric and Gas, Scottish and Southern Energy. These reports were selected for the reasonably comprehensive coverage of the range of issues that they give. (Fuller details of their description of effects are given in the Appendix.)

Table 2.1 shows a number of principal issues:

- Pollution: the control of emissions from equipment, especially of oil into ground and water;
- Climate change: the need to decrease greenhouse gas emissions, especially sulphur hexafluoride from substation equipment;
- Resources: the minimisation of resource use and wastage, especially through reduced energy consumption and waste management;
- Landscape: the reduction of the visual impact of overhead lines and towers;
- Conservation: the prevention of harm to wildlife and the cultural heritage;
- Human health and nuisance: the need to allay fears about EMF, and the minimisation of noise.

Table 2.1 Environmental effects of HVDT considered in corporate environmental reports (see Appendix for fuller details)

ENVIRONMENTAL EFFECTS	RESPONSES TO EFFECTS
Air and Climate	
Sulphur hexafluoride, used as an insulant at substations: a greenhouse gas; danger of leaks from equipment.	Losses recorded; measures to minimise.
Energy use (site-based): production of greenhouse gases.	Estimated emissions: to decrease.
Halon, used in fire-fighting equipment: ozone-depleting gas.	Incident causing release reported.
Ground and Water	
Oil, used as an insulant and coolant at substations: risk of leakage and spillage, causing ground and water pollution.	Pollution prevention measures; measures to deal with existing contamination.
Polychlorinated Biphenyls (PCBs), introduced into insulating oils: risk of oil leakage causing ground and water contamination, and take-up by living organisms.	Programmes of decontamination and removal of equipment.
Cable oil, used as an insulant in underground cables: risk of leakage, causing ground and water pollution.	Estimated leakages; reduced through monitoring and maintenance; research into means of minimising.
Pesticides, used to control vegetation: risk to wildlife.	Estimate of use.
Land	
Disturbance to land through construction and maintenance.	Ensuring minimum disturbance.
Non-operational land: possible harm.	Maintenance standards set.
Landscape	
Overhead lines and towers: visual intrusion on the landscape.	Improved routeing and design, care taken in designated areas, possible use of undergrounding.
Habitats and Wildlife	
Bird collisions with overhead lines.	Methods of preventing.
Overhead line corridors: visual effect of cutting vegetation.	Softening of edges by allowing shrub growth.
Disturbance to sensitive areas by construction work.	Consultation with conservation bodies to avoid harm.
Heritage	
Disturbance to archaeological sites by construction work.	Consultation with archaeological bodies to avoid harm.
Human Health and Nuisance	
EMF, emitted by conductors: perceived risk to health.	Recognition of public fears, although research has not established any risk; advice offered to the public.
Noise, generated by substation equipment and overhead lines: nuisance to residents.	Keep to a minimum; methods of reducing noise.
Resources	
Water consumption by plant operations.	Aim to reduce water usage.
Energy use by plant operations.	Aim to reduce energy consumption.
Energy losses in transmission of electricity.	Reduction in losses through equipment improvements.
Waste production through operations.	Waste management strategies to minimise waste.

These issues suggest that there are a number of pressures upon the utilities motivating concern with 'environmental performance'. Firstly, the reports display a response to pollution control legislation, most overtly to that limiting polychlorinated biphenyl (PCB) : concentrations in equipment. Similarly, specific treatment of climate change by one utility is a direct response to a government call for companies to report on their emissions (NG 1999b). Secondly, there is a clear environmental management ethos expressed in the reports; this is most evident in relation to quantifiable emissions, where targets can be set for their reduction, and in the auditing of resource use and waste management. The three companies concerned are all working towards implementation of a formal environmental management system, ISO 14001 (International Organisation for Standardisation 1996). Thirdly, the inclusion of landscape and conservation issues appears to be a response to land-use planning requirements, such as Schedule 9 responsibilities. These are generally treated in a more peripheral manner than the issues relating to pollution control and resource use.

These reports are therefore dominated by those effects of operations that involve quantifiable flows of materials and emissions, and keep in mind compliance with pollution control legislation and the targets of environmental management systems. Effects that fall more into the arena of land-use planning are given less importance; however, these 'softer' issues are those where public concern may be greater. The tone of the reports is an indication of their source, as they tend to be the responsibility of environmental management, rather than planning, staff. Given that reports of this kind can be valuable for communicating with a wide audience (Hunt and Johnson 1995), there is potential for giving planning issues greater weight within them.

2.9 ENVIRONMENTAL STATEMENTS FOR PROPOSED PROJECTS

According to European legislation, EIA "shall identify, describe and assess ... the direct and indirect effects of a project" (EC Directive 97/11, Article 3). Environmental statements and other reports produced by electricity utilities, whether under EIA requirements or as a form of good practice (2.5), therefore reveal the industry's perspective on the environmental effects of its activities.

NG, in preparing EIA documents for overhead lines, recognises a 'standard range' of issues that are likely to be covered by assessment (NG 2000):

- landscape and visual impact;
- ecology and water quality impacts;
- noise impacts;
- archaeological impacts;
- EMF;
- electromagnetic compatibility;
- construction impacts;
- operational impacts.

Table 2.2 shows the environmental effects considered in eight recent EIA documents prepared by NG for a range of transmission projects, covering the main components of the transmission system. (Fuller details of the effects considered are given in the Appendix.) This table shows a slightly fuller list of effects than the 'standard range'.

Table 2.2 Environmental effects considered in National Grid environmental statements and reports (see the Appendix for fuller details)

ENVIRONMENTAL EFFECTS	
Landscape and Visual	
Landscape and visual	Effects on landscape and views
Ground and Water	
Ground	Effects on soil and rock
Water quality	Effects on surface and ground water quality
Drainage	Effects on surface and ground water levels and movement
Habitats and Wildlife	
Flora and fauna	Effects on habitats and their species, especially on those of conservation value
Birdstrike	Likelihood of birds colliding with overhead lines
Cultural Heritage	
Archaeology	Effects on features of archaeological importance
History and listed buildings	Effects on sites of historical importance and on listed buildings
Human Health and Nuisance	
EMF	Effects of EMF on human health
Electromagnetic Compatibility	Effects of electricity transmission on television, radio and computer equipment
Noise	Nuisance effects of noise
Land use and Transport	
Agriculture	Effects on agriculture
Residential	Effects on existing or future residential land use; property prices
Recreation and tourism	Effects on recreation and tourism
Industry	Effects on industry
Minerals and waste	Effects on existing or future mineral extraction or waste disposal
Transport	Effects on roads, railways, paths, etc. and on movement of traffic, aircraft, etc.

Comparing Table 2.2 with Table 2.1 shows a difference of emphasis in the issues covered by corporate environmental reports and EIA documents. Measured against the corporate reports, the EIA documents, in general:

- do not consider questions of resource use and waste;
- cover some, but not all, of the potentially polluting effects of installations;
- give more comprehensive coverage of 'amenity';
- extend the scope of issues to include economic and social issues.

These differences are partly because EIA is concerned with individual projects rather than the overall operations of a company; this makes it possible, for example, to consider other immediate land uses and activities. The omissions in relation to resource use and emissions are perhaps surprising in the light of the legislative requirements for EIA to provide comprehensive coverage of effects; a broader scope of impact study might be expected. However, Table 2.2 does suggest that EIA legislation (2.5) is driving a more thorough analysis of the environmental consequences of HVDT than has been brought about by the Amenity Clause. It is possible, though, that the historic importance of the Amenity Clause in the industry has contributed to the place given to 'amenity' issues in EIA, having helped to shape the professional culture within which EIA is carried out. It is also likely that the emphasis within EIA on external consultation and public participation is a factor in determining the areas to which prominence is given; this could favour the inclusion of issues of greater public concern than those included in corporate environmental reports.

2.10 INDUSTRY PLANNING GUIDELINES

Proposals for new HVDT installations are generally put forward with regard to the industry's own planning policy and guidelines. These too reveal how the industry perceives the possible effects of its operations. There are two bodies of documentation in this respect:

- Schedule 9 statements, prepared under the Electricity Act, stating how licence holders intend to fulfil the obligations of the Amenity Clause (2.3.3).
- Voluntary planning guidelines prepared by individual utilities.

2.10.1 Schedule 9 Statements

Unsurprisingly, Schedule 9 statements echo the concerns of the Amenity Clause itself, but also outline some of the environmental priorities identified by the electricity utilities when planning new installations for which consent is required under the Act. For example, the statement of Yorkshire Electricity asserts, using language similar to that of the Amenity Clause:

> In all our new building and construction proposals ... we shall endeavour to be non-obtrusive insofar as this is reasonably practicable. In particular, within reasonable economic grounds, we shall take into account and try to mitigate the effect of such proposals on:

- the natural beauty of the countryside;
- flora, fauna and sites of geological and physical interest;
- all features, buildings, sites and structures of special architectural, archaeological and historical interest (Yorkshire Electricity 1999).

Some statements show particular regard for sensitive and designated areas. South Wales Electricity (undated) states, "In considering proposals for new developments we will consider whether we can exclude them from areas of high amenity value". Similarly, NG (2001a) states that, in formulating any relevant proposals, it will seek to avoid areas designated nationally or internationally for landscape, wildlife or cultural reasons.

Some statements specify possible means of avoiding or lessening the impact of installations. For example, South Western Electricity (undated) refers to the careful routeing of OHLs in order to minimise their visual impact, the screening and landscaping of substations, the reduction of equipment size, and the possibility of undergrounding lines in sensitive locations.

As these statements are drawn up primarily with the aim of satisfying the requirements of the Amenity Clause, they concentrate on those effects of HVDT installations that impinge upon the conservationist concerns of the Clause. Particular emphasis is placed upon the physical presence and visual appearance of equipment, particularly with regard to designated areas and buildings. Implicit in the statements is an acknowledgement of the visual intrusion of HVDT equipment.

2.10.2 Voluntary Planning Guidelines and Advocacy Documents

A utility's internal guidance indicates its environmental priorities in the design and planning of new installations. A number of NG documents illustrate guidelines of this kind; these take the form of internal notes, and publications designed for external distribution advocating NG's approach to the planning of new equipment. Table 2.3 summarises the effects that are referred to in these documents, and the measures outlined to mitigate these effects.

These documents emphasise the visual impact of new installations, and put forward certain mitigation measures. With regard to OHLs, careful line routeing is suggested. Reference is made to 'The Holford Rules', a set of seven guidelines for the least visually disruptive siting of OHLs. These were formulated in 1959 by Lord Holford, a member of the then Central Electricity Generating Board, and are still considered to provide a basis for line routeing. One of the 'rules' reads:

> Choose hill and tree backgrounds in preference to sky backgrounds wherever possible; and when the line has to cross a ridge, secure this opaque background as long as possible and cross obliquely when a dip in the ridge provides an opportunity … (NG unpublished.a).

Table 2.3 Environmental effects considered in National Grid planning guidelines

ENVIRONMENTAL EFFECTS	MITIGATING MEASURES
Guidelines for Line Routeing (NG unpublished.a)	
Visual impact of OHLs	Holford Rules – effective routeing of OHLs: avoiding designated areas, direct routes, having regard to backgrounds, avoiding concentrations of lines, etc.
Substations and the Environment (NG unpublished.b)	
Possible damage to sites of amenity, cultural, etc. value	Avoid designated areas of the highest value
Visual impact of substations	Screening; low profile designs; lightweight materials; smaller, equipment; separation of OHL entries
Equipment noise, security lighting	Attenuation measures
Water pollution due to leakages; land drainage effects	Pollution prevention measures
Sterilisation of land; effects on surrounding land use	
Undergrounding of High-Voltage Cables (NG unpublished.c) and **Overhead or Underground?** (NG unpublished.d)	
Excavation for installation and maintenance	
Possible damage to archaeological, etc. sites	
Land-take for trenches and compounds	
Severely restricted land use above cable route, including agricultural activity and tree growth	
Visual impact of sealing-end compounds and other equipment	
Noise from underground cable equipment	
Water pollution from underground cable fluids	
Development Near Lines (NG 2001b)	
Visual impact of OHLs	Alternative routes; sensitivity to landscape and designated areas; avoid residential areas; design of nearby development to minimise impact
Visual impact of substations	Substation design; landscaping; design of nearby development to minimise impact
Limitation of development near installations (safety and maintenance)	Consultation with NG regarding development near existing installations
Noise produced by installations	Attenuation measures; design of nearby development to minimise impact
Electric and magnetic fields (EMF) from equipment	Public concern recognised, though no scientific basis recognised; provide information
Other electrical effects: micro-shocks near OHLs; computer screen interference; electromagnetic compatibility issues	Earthing of metal objects near OHLs; equipment design; aerial siting
Safety considerations	Security features on installations

These guidelines reflect a tradition within the industry of attempting to minimise the effects of OHLs on the landscape, and are in keeping with the general tenor of the Amenity Clause. Equivalent guidelines are also set out for the siting and design of substations (NG unpublished.b).

Two of the documents respond to calls for OHLs to be laid underground as a means of minimising their visual impact. They argue that the effects of undergrounding are considerable in themselves, because of the major excavation required, for instance. These considerations, along with the far higher costs involved, provide the basis for NG's policy on undergrounding, which is to reserve it for exceptional locations, such as in some urban areas and designated areas; otherwise, NG aims to make "overhead connections wherever possible" (NG unpublished.c). One of these documents, *Overhead or Underground?* (NG unpublished.d) is a booklet published for external distribution in order to persuade readers of the considerable difficulties involved in undergrounding.

It is noticeable, however, that these documents make no mention of possible means of minimising the effects of undergrounding; this appears to be an expression of the strong policy stance taken on the issue and, in the case of the booklet, its persuasive purpose.

Development Near Lines (NG 2001b) is a publication aimed at LPAs and developers dealing with issues most likely to arise within the context of local planning. Amongst other things, it refers to the visual and noise effects that can be caused when new development takes place close to existing OHLs and substations. The booklet then lays stress on the layout and design of nearby development as a means of minimising these impacts, thus arguing that amenity considerations are the responsibility of developers as well as the NG. It also points out the difficulties for access and maintenance of OHLs that can be created by other development. The understanding of environmental effects is therefore expanded to include possible future effects caused, or at least intensified, by subsequent developments.

This booklet also discusses the issue of health fears associated with EMF. The findings of scientific research are quoted to support the assertion that there is no proven risk. In particular, the findings of the National Radiological Protection Board (NRPB) are relied upon as the basis for planning decisions, the conclusion being that "scientific evidence does not support any restriction on development in the vicinity of electricity transmission lines ..." (ibid. p. 16). The only 'effect' that is recognised is that of public concern over the issue, to which the response is to provide information and advice on relevant research.

These documents illustrate the environmental perspective of the electricity industry's planning section. The selection of issues dealt with is shaped by the requirements of land-use planning, focusing on 'amenity' concerns; the pollution and resource issues covered in corporate environmental reports are largely omitted. However, it could be argued that certain pollution issues are relevant considerations for planning, in that they can affect surrounding land uses. Also, there is scope in the planning guidance for more comprehensive coverage of effects of the kind presented in environmental statements. It appears that the coverage of effects in these documents is largely a reaction to the issues raised by LPAs, and can therefore be interpreted as the industry's understanding of and engagement with LPA concerns. This is most visibly the case with regard to EMF, which the industry finds itself obliged to regard as an issue despite its own scientifically based conclusion that there is no associated health risk.

2.11 CIGRE PAPERS

CIGRE gathers together the research findings and technical innovations of HVDT experts across the world, through conferences, published papers, and so on (CIGRE 2000). Although CIGRE's emphasis is on electrical engineering aspects of HVDT, the handling of environmental effects is implicit in much of its work; specific attention is also given to planning aspects of HVDT. As an illustration of the environmental issues raised, relevant papers presented at symposia are examined here. Although these papers are international in scope, they provide a complementary industry-based perspective to that of the UK sources reviewed above.

Many of the papers report increasing public opposition to HVDT installations, and some refer to the difficulties that this presents to utilities in gaining consent for new projects. This is exemplified by the increasing length of time required for the planning phase of a project (Bohme *et al.* 1998). Practical difficulties of this kind have played a part in making electricity utilities identify and address environmental aspects of HVDT, and seek technical means of mitigation. The effects and associated mitigating measures referred to in the papers studied are summarised in Table 2.4.

The visual appearance of OHLs, and to some extent substations, occupies the attention of many of the papers, and provokes a range of mitigation measures. For example, a major project to renew the Danish transmission system has been undertaken as a direct response to public reaction against the appearance of OHLs and a fall in the value of nearby property (Engsbro *et al.* 1996). The programme involves a combination of measures: routeing lines away from densely built-up areas and protected areas; undergrounding lines; the use of steel poles instead of lattice towers; and the use of towers to support circuits of different voltages. This illustrates a change in emphasis from satisfying purely engineering criteria in the design of lines to an attempt to minimise their environmental impact (Luain and Corcoran 1996).

Specific attention is given to the undergrounding of lines, as a less intrusive, though costly, alternative to OHLs. Locations where undergrounding might be considered include those where OHLs would be visible over large distances, or where there are strong EMF concerns (Bascom *et al.* 1996). But undergrounding has its own impacts, which, along with its high costs, impose severe constraints on its applicability. These effects are looked at in a study undertaken by CIGRE (CIGRE Joint Working Group 1996). However, this study is comparative, setting out the effects of undergrounding alongside those of OHLs, as NG has done (2.10.2).

Table 2.4 Environmental effects of HVDT considered in CIGRE papers

ENVIRONMENTAL EFFECTS	MITIGATING MEASURES		AUTHORS
Visual Effects and Overhead Line Routeing			
Visual impact of OHLs and substations	Equipment designed to be less visually intrusive	Towers: new designs	Luain and Corcoran (1996)
		'Compact' overhead lines	Ojala *et al.* (1998), Bohme *et al.* (1998)
		Camouflaging of towers and conductors	Egger and Lugschitz (1998)
		Substations: more sensitive design and screening	Adema *et al.* (1996), CIGRE Joint Working Group (1998)
	Routeing of OHLs to take greater account of surrounding landscape, and to avoid sensitive areas	Landscape survey and visualisation of proposed lines: advanced techniques	D'Adjello *et al.* (1996), Doyen-Viguier *et al.* (1996) Luain and Corcoran (1996), Rogier *et al.* (1996), Toshirou *et al.* (1996)
		Greater consultation in choosing line routes	Doyen-Viguier *et al.* (1996), Fitzgerald *et al.* (1996), Kim *et al.* (1996), Lomas *et al.* (1996)
	Undergrounding of OHLs	Consideration of undergrounding in sensitive locations	Bascom *et al.* (1996), CIGRE Joint Working Group (1996), Engsbro *et al.* (1996)
	Multiple solutions, making use of a number of measures		Engsbro *et al.* (1996), Luain and Corcoran (1996)
Undergrounding			
Adverse effects of undergrounding: restriction on land use, land and water pollution	Comparison of effects of OHLs and underground cables		CIGRE Joint Working Group (1996)
EMF			
Perception of health risks			Engsbro *et al.* (1996), Henning *et al.* (1996), Luain and Corcoran (1996)
(No environmental effects defined)	'Compact' OHLs; different configurations of conductors		Alexandrov *et al.* (1998), Awad *et al.* (1998), Kikuchi *et al.* (1998), Loudon *et al.* (1998), Ojala *et al.* (1998)
Micro-shocks, interference with computer screens	'Compact' OHLs		Behncke *et al.* (1998)
Noise			
Mainly associated with OHLs under certain weather conditions	Line routeing to increase the distance between the noise source and the receptors		Lomas *et al.* (1996)
	Conductors designed to emit minimal noise		Bohme *et al.* (1998), Kim *et al.* (1996)

The reduction of EMF emitted by OHLs is a major preoccupation of the papers, and a number of engineering solutions are proposed to this end. It is not always clear, however, which effects of EMF are being countered. The fear of health risks associated with EMF is recognised as a cause of opposition to new installations (Engsbro *et al.* 1996, Luain and Corcoran 1996), and in one case, proposals for a 'compact' line are made specifically with health anxieties in mind (Henning *et al.* 1996); this is despite the industry's assertion that there is no proven health risk resulting from EMF exposure (2.10.2). It appears that, as in EIA practice (2.9), the need to address public fears overrides the industry's view of scientific evidence. Here, this need extends beyond the content of environmental statements to the design of installations, possibly with considerable cost implications. However, other less debatable effects of EMF are also likely to be in mind, such as the possibility of micro-shocks and interference with computer screens (Behncke *et al.* 1998).

Comparing Table 2.4 with Table 2.3 shows that these papers concentrate on similar issues to those raised in NG's planning guidelines, that is, those of most immediate concern to local planning. A relatively narrow set of preoccupations surfaces consistently in different countries, especially in more industrially developed nations. However, the international literature suggests that a more active approach is being taken outside the United Kingdom to mitigating measures. Greater environmental acceptability of HVDT installations is being sought not just by the industry's planning sections confronted by the increasing difficulties of gaining consents, but also by engineering sections responsible for the design of equipment; substantial technical developments are being advanced that specifically aim to mitigate adverse effects. Bohme *et al.* (1998) claim that an 'environment friendly' OHL can only be achieved through new design concepts of this kind.

2.12 ELECTRICITY INDUSTRY PERSPECTIVES ON THE EFFECTS OF HVDT

This review reveals different perspectives on the environmental effects of HVDT activity within the electricity industry. Varying emphases reflect different responsibilities within the industry, and responsiveness to a range of legislative duties, business practices, public pressures and technical possibilities. There is, nevertheless, considerable common ground between the issues identified by different sections of the industry, and a high degree of consensus regarding the industry's attitudes towards those issues. Figure 2.1 shows a synthesis of the effects described in the documents reviewed above. This diagram summarises the overlapping perspectives within different sections of the industry on the interaction between HVDT and the surrounding environment.

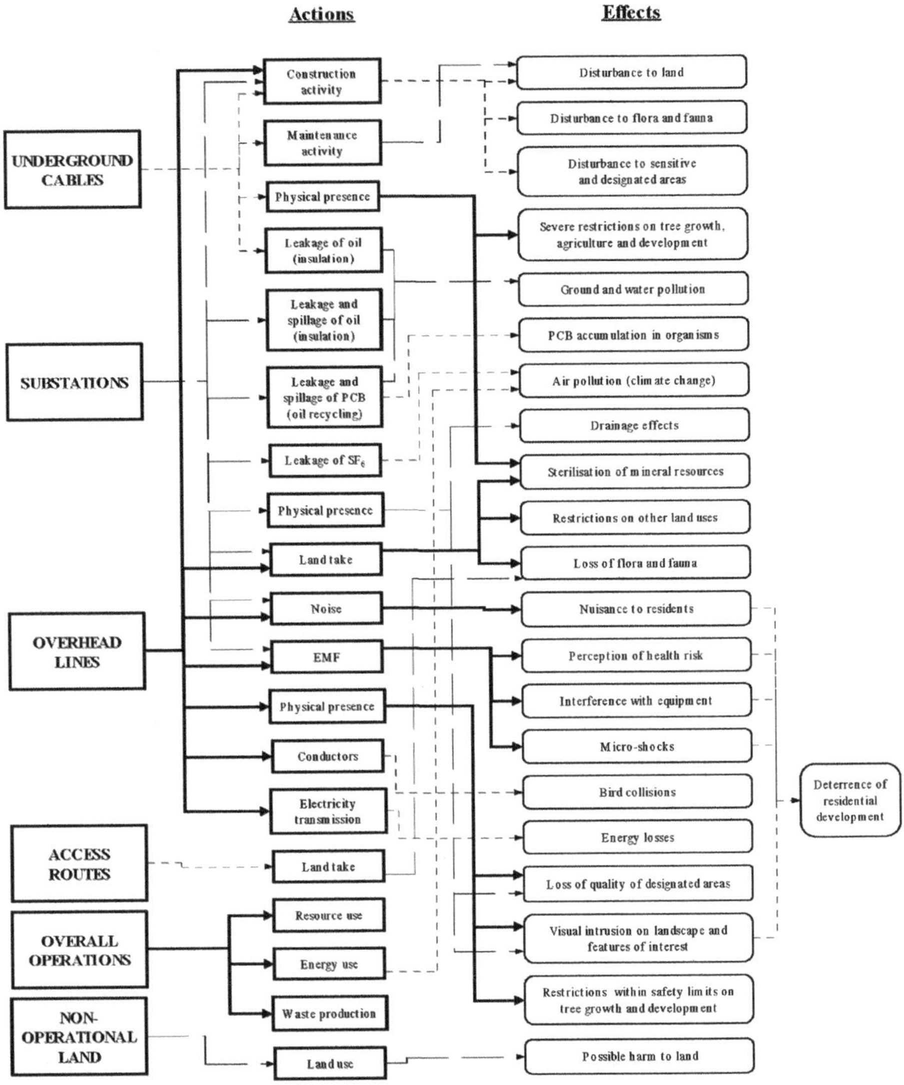

Figure 2.1 Environmental effects of HVDT described by the electricity industry

2.13 CONCLUSION

This overview of the legislative framework within which the planning of HVDT installations operates, and of the industry's understanding of the environmental effects of its operations, has raised a number of issues of potential importance to local authority HVDT-related policy. These can be summarised as follows:

- The consents procedures for HVDT installations place considerable emphasis on environmental concerns, through the Amenity Clause and EIA regulations; however, other duties on licence holders may exert pressures not always compatible with environmental responsibilities.
- LPAs have a limited, but influential role in the planning of HVDT installations, which may encourage electricity utilities to enter into negotiations with them.
- The electricity industry has a historic sensitivity to landscape, conservation, and so on issues due to the Amenity Clause; this still plays a large part in the industry's approach to planning matters.
- More recent EIA legislation and environmental management practice have driven a more rigorous and wide-ranging approach to the environmental effects of new installations.
- The industry has taken a defensive stance with regard to certain environmental matters, notably the undergrounding of lines and the possible effects of EMF.
- Although technical and other means of mitigating environmental effects are being pursued, UK initiatives in this regard appear to be less advanced than elsewhere in the world.

These issues will be returned to following the empirical study presented in Chapters 5 to 9. The next chapter, however, considers the procedural means by which HVDT-related policy is being advanced, the system of development plan preparation.

CHAPTER 3

The Development Plan System in England and Wales

3.1 INTRODUCTION

High-voltage distribution and transmission (HVDT)–related policy is not adopted by Local planning authorities (LPAs) as an independent statement of policy, but comes to expression through the preparation of statutory development plans (Section 1.3). The formation of HVDT-related policy is closely linked to the overall process of plan-making; it is likely to reflect the broader emphases apparent in development plans, and to be partly a product of the influences at work in plan preparation as a whole. Hence analysis of this policy area requires an understanding of the wider process of development plan preparation, and the trends that are currently apparent within this activity. This chapter explores the nature of plan-making in England and Wales, drawing out features of particular relevance to the formation of HVDT-related policy.

The chapter considers, firstly, the establishment of the system of development plans in force at the time of the research (this is currently undergoing reform, under the Planning and Compulsory Purchase Act 2004). Particular emphasis is placed upon the period since 1991, which has seen significant changes in plan-making; these include the growing importance of ideas of sustainability, and a heightened degree of public participation. This leads, secondly, into a consideration of the expression of interests and concepts in development plans. As different interests are, to some extent, mediated through the formal procedures of plan-making, the process of plan preparation is also discussed in this context.

Reference is made throughout this chapter, especially in the second half, to research and other academic work that explore recent trends in the preparation of development plans. This provides a further context within which the study of

HVDT-related policy can be placed, and allows the study of this particular policy area to contribute to a wider understanding of the processes currently at work in plan-making.

THE ESTABLISHMENT OF THE DEVELOPMENT PLAN SYSTEM

3.2 DEVELOPMENT PLANS FROM 1947 TO 1991

The Town and Country Planning Act 1947 established the principles and instruments of planning, which have provided the basis of the UK system to the present day (Rydin 1998). The Act included the establishment of statutory development plans, to be prepared by county and borough councils, which were to indicate "the manner in which they propose that land in their area should be used" (Section 5). Two characteristics of development plans were introduced at the outset. Firstly, they were to deal with the physical use of land rather than broader social or economic issues. Current guidance continues to assert the same principle (DETR 1999a, para. 3.5). Secondly, plans were to provide a policy background for planning decisions, rather than to be prescriptive and legally binding in effect; again, plans still retain this discretionary nature (Hughes 1997).

Development plans have been subject to various revisions since their inception, usually in response to difficulties encountered in preparing and implementing them (Moore 2000). The year 1968 saw a major change, with the introduction of a two-tier system: this required county councils to prepare structure plans (SPs) providing a strategic planning framework, and district councils to prepare local plans (LPs) providing more detailed guidance. Other modifications were made in the 1970s, though these had the effect of downgrading plans, by stating, for example, that LPs need not be prepared for the whole of a district's area (Davies 1999). In addition, the practice of plan-making proved to be a lengthy and laborious process, with numerous, overly detailed policies being advanced; nonstatutory planning documents also proliferated, adding to a sense of uncertainty about the role of plans (Cullingworth and Nadin 2002).

These concerns were added to in the 1980s by criticism levelled at the planning system in general. The political climate during the Thatcher-led Conservative government was one of deregulation, in which planning procedures were seen as holding back development. Development plans, especially SPs, came under attack for being too detailed and complex in content, and too cumbersome and slow to produce (Davies 1999). This resulted in the relegation of plans to being "one, but only one, of the material considerations that must be taken into account in dealing with planning applications" (DoE 1985, para. 5). However, this period also saw the introduction of unitary development plans (UDPs), containing sections equivalent to SPs and LPs (DoE 1992a); this was necessitated by the creation of single-tier metropolitan authorities following the abolition of metropolitan councils.

3.3 DEVELOPMENT PLANS SINCE 1991

3.3.1 The Planning and Compensation Act 1991

In view of the effective downgrading of development plans that had taken place in the 1980s, the 1990s saw a surprising upturn in their importance. This followed in the wake of increasing unease about the consequences of weakening planning controls. Not only were local authorities objecting to the undermining of their position, but a "presumption in favour of development" had encouraged what was seen as unbridled development, and was arousing opposition from many Conservative voters and members of Parliament, especially in southern England (Tewdwr-Jones 1996).

It is within this context that legislation for the future of the planning system was being prepared. The resulting Planning and Compensation Act 1991 dealt specifically with development plans, and stated that planning decisions were to be made "in accordance with the plan unless material decisions indicate otherwise" (introduced as Section 54A into the Town and Country Planning Act 1990). This came to be described as a "plan-led system" (a phrase coined by the Minister for Housing at the time). Other key changes affecting the role and preparation of development plans were introduced; these included the requirement for district councils to prepare LPs for the whole of their areas, and procedures designed to speed up plan preparation (Tewdwr-Jones 1996). This legislation was closely followed by a government "planning policy guidance" (PPG) note to LPAs on the preparation of development plans, stating their importance to local planning (*PPG* 12):

> Development plans set out the main considerations on which planning applications are decided and can guide a range of other responsibilities of local government and other agencies (DoE 1992a, para. 1.4).

It appeared, on the face of it, that development plans had experienced a reversal of fortune, and were now at the very heart of the planning system.

3.3.2 The "plan-led system"

There is debate, however, over how real a change was brought about by Section 54A (Adams 1994). Firstly, the Act itself qualified the primacy of plans, by stating that planning decisions are to be made "in accordance with the plan *unless material considerations indicate otherwise*" (emphasis added). Secondly, other government guidance, *PPG 1*, published in the same year as *PPG 12*, continued to assert unambiguously the presumption in favour of development, saying that the planning system "should operate on the basis that applications should be allowed, having regard to the development plan and all material considerations ..." (DoE 1992b, para. 5). Rather than the development plan being the primary factor in planning decisions, other factors retained the possibility of being decisive. Indeed, a number of judicial decisions on Section 54A have weighed in favour of other material considerations (Gatenby and Williams 1996). Further ambiguity is caused by the absence of any definition of these other considerations, and uncertainty about the relative degree of weighting that should be attached to them (Tewdwr-Jones 1994a).

A particular difficulty is raised by the relationship between development plans and national policy. The late 1980s and early 1990s saw the publication of a whole series of PPG notes, forming a more systematic body of policy at the national level than their precursor circulars had provided. This created the potential for conflict between these notes and development plan policy. Tewdwr-Jones (1994a) studied the degree to which development plans were standing up against national guidance at appeals; this revealed considerable ambiguity over the status that each should hold. This uncertainty is to some extent resolved by a hierarchical view of policy, in which development plans conform with national guidance, but this then weakens the ability of plans to reflect local conditions (Tewdwr-Jones 1994b).

The supposed primacy of development plans following the Planning and Compensation Act has therefore been thrown into doubt. Indeed, from a legal perspective, Section 54A appears to have made little difference to the status of development plans (Gatenby and Williams 1996), except that they themselves may now be a stronger material for consideration than previously. Rather than there being an unambiguous "plan-led system", it may be more accurate to say that there is now a presumption in favour of development in accordance with the plan (Edwards and McCafferty 1992).

3.3.3 Development plans in planning practice

Even if the "plan-led system" has a relatively weak legislative basis, there remains the question of the recent place of development plans within local planning practice. There seems to be little doubt that development plans have found a renewed importance in planning decisions since the Planning and Compensation Act. For example, in a study of areas under considerable pressure for development, Hull and Vigar (1998a) found that plan-making had become more central to the activities of LPAs. It had assisted in the management of growth throughout their areas, partly by creating an opportunity for different interests to be expressed, and partly by providing policies upon which LPAs can base a defence of their decisions at appeals.

Similarly, MacGregor and Ross (1995) noted an increase in the practical role of development plans since 1991. However, they did not attribute this to the introduction of Section 54A, but to a number of other possible reasons:

- other expressions of government policy appear to enhance the status of development plans;
- LPAs are more likely to be supported on appeal if their decisions are based on written, local policy;

- the Planning and Compensation Act has made LPs mandatory;
- the move towards sustainable development requires more substantive plan policies.

They asserted that the increased status of plans was one of perception rather than legal standing; this was particularly so amongst groups that were taking an increasingly active role in plan-making, notably developers and community groups.

The popular planning press also provides an indication of greater importance being attached to development plans since 1991, particularly by planning officers. Here, there has been a strong feeling that Section 54A introduced a new emphasis, and the rhetoric of the "plan-led system" has been prevalent (e.g. Baker 1997, Deasley and Hamilton 1995). This is matched by continuing reports and discussions about the performance of plans and the difficulties encountered in plan preparation, reflecting a renewed surge of activity in plan-making (e.g. Baber 1996, Milne 1997). Particular concern has been expressed about the length of time taken in bringing plans to adoption, largely because of the vastly increased number of objections being made to proposed policies during consultation stages (e.g. Crouch 1995, Fair 1994).

It is clear, therefore, that development plans have been established more firmly within the planning system since 1991 than previously. This owes a certain amount to the Planning and Compensation Act; in particular, the requirement for plans to provide universal local-level coverage initiated the preparation of a new generation of plans. Beyond this, the legislative effect of the Act has been relatively weak and cannot be said to account for the revival of development plans. However, Section 54A has had a significant indirect influence, in that it has shaped national policy on development plans as expressed in *PPG12*, and has also created a widespread perception that plan-making is a key element of the planning system.

3.3.4 Development plans and sustainability

As mentioned above, MacGregor and Ross (1995) suggested that the increased status of plans since 1991 was connected with a simultaneous emphasis upon sustainable development. This was initiated at the government level with the publication of the white paper *This Common Inheritance* (Her Majesty's Government 1990), which marked a clear shift to a new environmental consciousness (Owens 1994). Indeed, the Planning and Compensation Act, following closely behind, may have been drafted under the assumption that development plans would be a means of achieving greater environmental protection. Certainly, *PPG 12* described the "plan-led system" along these lines:

> The planning system, and the preparation of development plans in particular, can contribute to the objectives of ensuring that development and growth are sustainable (DoE 1992a, para. 1.8).

This PPG also recommended that plans should be subject to environmental appraisal, further strengthening the place that environmental concerns should have within them. These emphases were reasserted in a revision of *PPG 12* published by the new Labour administration in 1999. Here, development plans were linked even more strongly to the government's strategy for sustainable development, although sustainability was explicitly stated to encompass social and economic, as well as environmental, aims (DETR 1999a).

Government guidance since 1991 has therefore attributed to development plans the specific purpose of contributing to the achievement of sustainable development. As a result, the fortunes of development plans have become linked to the importance of sustainable development in public policy. "Sustainable development", or "sustainability", has, in fact, taken on an overriding importance during this period. Despite difficulties of definition, it has come to suffuse policy at many levels, such that it can be described as "a term over which there is almost universal consensus ... the postmodern equivalent of a grand narrative" (Myerson and Rydin 1996, p. 22–23). Certain principles of sustainability are becoming more integrated into planning as a whole (Wilson 1998). Development plans themselves have increasingly exhibited this trend, with greater use of the language of sustainability evident throughout their texts. Whether plans are effective tools for pursuing sustainability objectives is open to question; but the notion that they are appears to have contributed significantly to their new-found role since 1991.

3.3.5 Public participation in plan-making

The statutory procedures for development plan preparation have included requirements for public participation since 1968; indeed, the process of plan preparation is largely dictated by these requirements (Figure 3.1). However, formal participation has been confined to opportunities for the public to make representations on proposed policies rather than to participate in the initial formation of policy. Although some LPAs have sought to engage communities more openly in policy creation, the emphasis has generally been upon LPAs presenting their proposed plans, and then awaiting public reaction (Rydin 1999).

Plan-making since 1991 has been characterised by a greatly heightened level of this form of public participation. The number of representations, mostly objections, being made on proposed policies at the early stages of plan preparation has escalated, often leading to lengthy and confrontational public inquiries. For example, one LP prepared in the late 1980s raised just 350 objections at deposit stage, leading to a public inquiry lasting four weeks; whilst a revision of the same plan just a few years later, in the early 1990s, raised 1500 objections, leading to an inquiry lasting 22 weeks (Crouch 1995). Typically, objections are brought both by developers objecting to constraints imposed by plans, and pressure groups opposing proposed development sites. The different interests that have become involved in this way are discussed further below. The fact that they are choosing to press their cases through the process of plan-making is significant in itself; however, as it suggests that they consider development plans to be important and relevant to planning decisions, their attempts to influence plan content are warranted. Indeed, the introduction of the "plan-led system" can be interpreted as a signal to stakeholders (especially developers) that plan preparation, rather than development control, is now the arena in which to try to bring their influence to bear (Hull and Vigar 1998b).

3.3.6 A central place for development plans

Development plans have therefore taken a far more central place in the planning system since the Planning and Compensation Act. Although the Act did little to give greater legal weight to plans, it has had far-reaching practical implications, especially because of its requirement for comprehensive, detailed plan coverage of England and Wales. This, along with recently established national and regional planning guidance, encouraged Davies (1999, p. 58) to conclude that "by the fiftieth anniversary of the passing of the 1947 Town and Country Planning Act there would be in place, or near to completion, an intelligible, comprehensive and effective hierarchy of plans". Plan-making has therefore become a renewed and essential activity for LPAs, gaining a strong institutional role.

Government policy has also ensured a new centrality for plans, by asserting their importance to planning decisions and to the achievement of sustainable development. PPGs have made specific reference to Section 54A, and have helped to establish the rhetoric of the "plan-led system". Even the ambiguity created by the original *PPG1* has been swept away in a revised version of the guidance, which states in clear terms, "The Government is committed to a plan-led system of development control" (DoE 1997a, para. 40).

3.3.7 The future of development plans?

Government commitment to a plan-led system has been confirmed recently in a planning green paper published in 2001. However, this paper also echoed the criticisms of plan-making made since the 1960s, suggesting that it is a laborious process resulting in complex and inflexible plans (DTLR 2001a). Major re-forms were proposed to overcome these difficulties, including the replacement of LPs, UDPs and SPs with single-tier "local development frameworks" containing core policies, followed by detailed action plans for areas where a change in land use is envisaged. These frameworks would be subject to sustainability appraisal, and also to a less time-consuming (though undefined) form of public participation. These proposals are now being implemented through the Planning and Compulsory Purchase Act 2004, and represent a significant change in the nature of plan-making in the United Kingdom, with potentially wide-ranging implications for the formation of local policy. For example, action plans may focus more attention on matters of immediate relevance to the areas in question. This is likely to result in greater weight being given to local conditions, such as environmental quality, but may well be at the expense of more strategic considerations, especially given the proposed abolition of plans at the county level.

Partly in response to the green paper, the Royal Commission on Environmental Pollution (RCEP) has argued for a continued strategic element within plan-making. In fact, the Commission advocates much wider ranging “integrated spatial strategies”, covering not just traditional land-use planning matters, but all spatially related activities, such as agriculture and forestry, pollution control, energy infrastructure, and so on. This is in the interest of “environmental sustainability”, whereby economic and social objectives take full account of environmental constraints (RCEP 2002). These recommendations are consistent with the linkage of development plans and sustainability, as noted above, but also attempt to reassert the primacy of environmental considerations in the sustainability equation. To what extent planning policy will develop in the directions suggested by the Commission remains to be seen.

THE EXPRESSION OF INTERESTS IN DEVELOPMENT PLANS

3.4 RECENT FEATURES OF PLAN-MAKING

The increasing importance of development plans since 1991 cannot be explained simply in legislative and policy terms. As shown above, their resurgence has been associated with two particularly strong themes, which are now explored further in order to provide a more detailed framework for the study of HVDT-related policy. These features, growing environmental awareness and greater public participation in plan-making, are themselves interrelated, and can be discussed in terms of the expression of different interests in plan-making. This discussion is prefaced by a consideration of the procedures laid down for public involvement in the preparation of development plans.

3.5 PROCEDURES FOR CONSULTATION AND PARTICIPATION IN PLAN-MAKING

The statutory procedures that an LPA must follow when preparing a development plan are explained in *PPG 12* (DoE 1992a, DETR 1999a). The main steps are summarised in Figure 3.1. The procedures make provision for the LPA to undertake early consultation with the public and certain bodies (including neighbouring LPAs, government departments, regulatory agencies and utilities) regarding their proposed plan. Although the form of consultation is not stipulated, and certain requirements have recently been lifted with respect to LPs and UDPs, the involvement of local people in plan-making is regarded as a key aim (DETR 1999a).

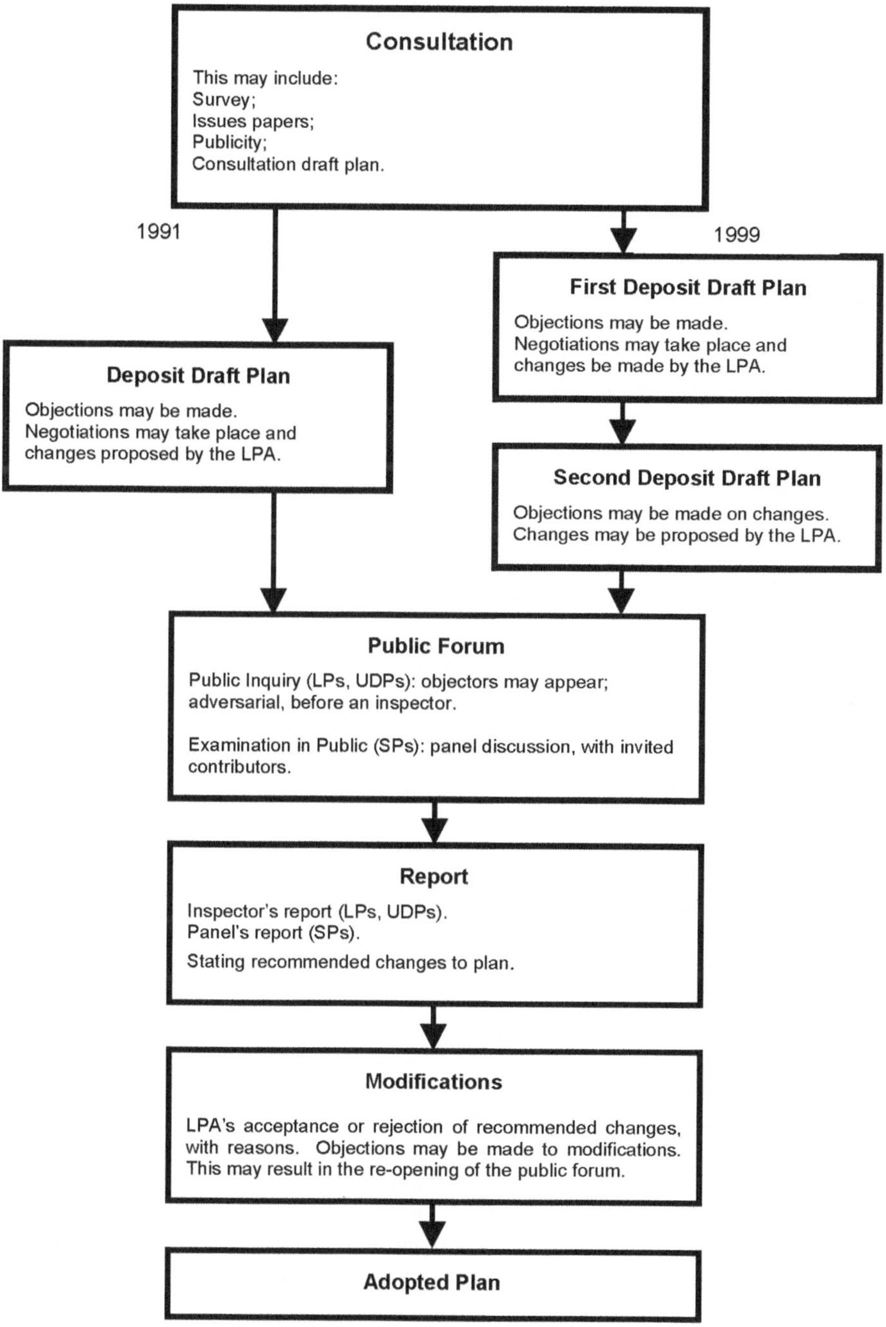

Figure 3.1 Main steps in the preparation of development plans

Extensive provision is also made throughout the procedures for objections to be made to proposed versions of a plan. These may be brought by individual members of the public, or by organisations, public, private or voluntary, and must be given consideration by the LPA. Objections can be made at three key stages: during early consultation, whilst the plan is placed on "deposit" and following final modifications to the plan (Hughes 1997). They may be dealt with by negotiation, or may proceed to a public forum; this takes the form of a public inquiry in the case of LPs and UDPs, or an examination in public in the case of SPs. When a public inquiry is held, objectors have the right to appear, and the inquiry typically takes on a quasi-judicial and adversarial nature, under the presidency of an inspector. By contrast, an examination in public consists of a panel, who invite certain representative objectors to participate in a round-table discussion.

The preparation of a development plan is therefore a highly structured process, consisting of a series of well-defined and regulated steps. Moreover, it contains many opportunities for the involvement of external bodies and individuals, through formal provisions for consultation and representation. Indeed, the steps involved in preparing a development plan are largely dictated by these provisions; engagement with external interests is integral to the entire process. It is partly the thoroughness of plan-making in this respect that has turned it into a long, drawn-out process, opening it to the criticisms and difficulties mentioned above (Sections 3.2, 3.3.5). Recent changes to the procedures attempted to reduce the time taken up by objections, especially in the preparation of LPs and UDPs (DETR 1999a), but the principle of extensive participation remains in place.

Plan-making is therefore formally accessible to those who seek to influence its content (Adams 1994). Policy is, in fact, likely to be considerably modified in response to the pressure of different, possibly competing, bodies at various stages during its formation; it will be shaped particularly by the interests that are expressed most strongly or successfully during plan-making.

3.6 THE ROLE OF DIFFERENT INTERESTS IN PLAN-MAKING

3.6.1 Promotion of interests in the planning system

Planning in the United Kingdom has often been understood in terms of mediating between competing demands for the limited resource of land. For example, Adams has described urban planning as a process of "bargaining, negotiation and compromise over the distribution of scarce environmental resources" (Adams 1994, p. 2). However, this does not necessarily result in an equitable sharing of resources; as Adams continues, "... the planning authority, in attempting to mediate between conflicting claims on land, may promote particular interests above others". Questions are therefore raised regarding the operation of different interests within the planning system and the degree of success of different groups in achieving their aims.

A number of studies have sought to investigate these questions. For instance, Blowers (1980) was concerned to determine the distribution of power in the formation of planning policy. He concluded that planning tends to reflect the existing pattern of power in society, and serves whichever interests are most influential at the time. Similarly, Healey *et al.* (1988) stated that planning "is tructured by the dominant power relations of society", which they identified as "the economic drive to secure profitability and the political conflicts over environmental quality and land values" (p. 244). Interestingly, they found that bias in the system was working in favour not just of industrial and commercial interests, but also of well-organised community and environmental groups, a trend that is still evident, as seen below.

3.6.2 Objections to proposed policies

There have also been studies paying specific attention to the role of different interests in the preparation of development plans. However, these indicate that plan-making has not always been a focus for the expression of conflicting interests. For example, a study by Healey in the early 1980s showed that negotiations between groups were taking place more within the context of development control than within policy-making (Healey 1983, p. 234). Equally, Bruton and Nicholson (1987), although they discussed the complexities and constraints faced by planners, did not indicate that plans were a particular focus for conflicting interests before the mid-1980s. This is likely to reflect the relatively weak status of plans, especially LPs, during this period (Section 3.2); participation in plan-making is unlikely to be strong if plans are perceived as having only a marginal role in planning decisions.

However, the expression of interests within the preparation and content of development plans took a new turn at the end of the 1980s, in tandem with their moving into the centre ground of planning (Section 3.3). Adams and Pawson (1991) carried out a study of representations made during the preparation of a sample of LPs during the late 1980s. They found that an increasing number of objections were being made to proposed policies by certain groups, in consistent and organised ways. Typically, these were being brought on the one hand by property companies favouring development, and on the other hand by residents' and amenity groups opposing proposed developments. Their study investigated the methods adopted by different groups in trying to achieve their aims; for instance, development interests tended to be professionally represented at inquiries, whilst residents' groups relied upon large numbers of written submissions. Unfortunately, the study did not attempt to analyse the underlying influences at work in the upward trend and the nature of these objections. It is likely, however, that they related to increased levels of entrepreneurial development and public environmental sensitivity throughout the 1980s, and that they were indicative of plans starting to become a renewed focus of planning activity.

3.6.3 Economic and environmental imperatives

A more interpretive approach to investigating the interests being expressed in plan-making was taken by Davoudi *et al.* (1996), who studied the preparation of the Lancashire SP in the mid-1990s. They identified two dominant objectives within the plan: economic development and environmental sustainability. These were each being pursued by clearly organised local alliances, involving businesses, voluntary and environmental groups, agencies, elected members, and so on. Tensions between these different interests persisted throughout the preparation of the plan and were evident in the final text. The different strategies adopted by the two groupings were seen by Davoudi *et al.* as expressions of wider forces, socio-economic and environmental respectively; for instance, environmental concerns were initially driven by an imperative to improve bathing water quality.

This analytical approach, of exploring the interests pursued in plan-making within their broader settings, provides a fuller understanding of underlying motivations of interests, and of the overall shaping of policy. This example also introduces the role played in policy formation by other actors, such as elected members and planning officers, not through the making of objections, but in the drafting of policy.

This case study of the Lancashire SP supports the assertion made by Healey (1994) that, in the 1990s, development plans took up a specific role of attempting to mediate conflict between environmental conservation and economic development. She saw this as a form of the established concept of planning handling competing demands, but emphasised that the interests in question were of a very contemporary nature. For example, the environmental agenda was being seen as encompassing biospheric as well as local issues, and the relation between differing interests was being expressed in terms of sustainable development. The attempt to find a balance between economic and environmental goals thus provided development plans with a new level of importance:

> In the ... 1990s, therefore, development plan-making has once again become a significant political arena within which struggles over the processes and purposes of local environmental change are being played out (Healey 1994, p. 48).

Certainly, there is strong evidence that, since the early 1990s, participation in plan-making has been targeted by environmental groups as a means of achieving their aims. This can be seen in a number of their publications, in which monitoring of existing policies is described, model policies are presented, and advice is given to supporters on how to influence plans being prepared locally (British Waterways 1991, CPRE 1992, Countryside Commission, English Heritage and English Nature 1996, Friends of the Earth 1994). For example:

> Local plans provide an ideal vehicle for putting concern for the environment into effect by means of policies for the use of land. The town and country planning system can be influenced by people with good arguments that are presented effectively (CPRE 1992, p. 3).

Equally, it is clear that some development interests are following plan-making closely in areas that are of concern to them (Section 3.3.5). To what extent these differing interests are being accommodated within the process of plan preparation is less certain. However, it does seem to be the case, as Healey suggested, that "sustainability" is increasingly being seen as the setting within which solutions to competing claims are being sought.

3.6.4 Notions of sustainability

A number of studies have attempted to assess the degree to which development plans have been incorporating notions of sustainability. For instance, Bruff and Wood (1995) measured the content of several UDPs from the early 1990s against a number of clearly defined sustainability principles; they concluded that only those principles that coincided with traditional planning concerns, such as to do with the built environment, were finding significant expression in the plans. A similar approach was taken by Counsell (1998, 1999) with regard to a sample of SPs from the mid-1990s; here, plans varied greatly in their inclusion of notions of sustainability, but plans prepared in the south-east of England, where pressure for development is greatest, performed highest in this regard. Although these studies did not identify any external pressures upon LPAs to prioritise sustainability objectives in plan-making, they did suggest that, in certain respects, sustainability notions are increasingly coming to the fore.

However, these studies were based on fixed assumptions about the understanding of sustainability; for instance, plans were assessed for their inclusion of clearly defined principles. But there is considerable debate about the definitions of sustainability being applied within planning. For example, Gibbs *et al.* (1998) described a spectrum of interpretations, from very weak to very strong sustainability, reflecting different opinions on the substitutability of resources. They concluded that local authorities have generally adopted weak versions of sustainability in their policy statements, resulting partly from tensions between economic and environmental goals. Similarly, Owens (1997a) noted that, although the rhetoric of sustainability has been widely accepted, divisions quickly appear over its implementation. In particular, she claims, there is a consistent tendency for environmental considerations to give way to economic "imperatives" (Owens 1997b). A similar assessment has been made more recently by Hales (2000). There is, therefore, pressure within plan-making for the adoption of a sustainability approach and the application of certain principles, but also considerable uncertainty regarding the precise nature of the concept, and there are serious difficulties in resisting the dominance of other perspectives.

Although the terminology of sustainability (and sustainable development) is relatively recent, having come to the fore via the Brundtland Report (World Commission on Environment and Development 1987), its appropriation within planning has, to some extent, relied upon existing planning concerns. For example, strong conservation traditions have been easily incorporated into notions of environmental management, or translated into ideas of environmental stock. Alongside this, however, has been the introduction of a broader dimension to planning, to give consideration to more global, or "biospheric", consequences of actions, and also to raise questions of limits to growth (Healey and Shaw 1993, 1994). It appears that the preparation of development plans, in particular, has provided an opportunity within planning for ideas contributing to "sustainability" both to evolve and to be introduced.

3.7 CONCLUSION

This chapter has traced the increased standing that development plans have gained in England and Wales, particularly since 1991, such that they now occupy a central place within planning practice. This establishment of a plan-led system, despite its ambiguities, has gone hand-in-hand with a heightened level of external involvement in plan-making. Members of the public, voluntary groups and commercial organisations have used the procedures of plan preparation to the full in order to express differing, often conflicting, interests. Their representations have brought to the fore widespread tensions between economic and environmental objectives, and have therefore been expressive of broader issues that are being addressed under the heading of sustainability. Plan-making is, in fact, seen as an important arena for debating questions of sustainability, and owes some of its increased status to this perception. The "strength" of the sustainability principles that are being incorporated into development plans remains doubtful, however, reflecting continued struggles between differing perspectives in the planning realm. The future direction of development plans also remains uncertain, following publication of the planning green paper.

This assessment of the current status of development plans, and of the trends apparent in their preparation, provides a framework within which to study the topic of HVDT-related policy. It raises questions about the formation of this policy area: Has HVDT-related policy emerged out of the revival of plan-making? What role do differing interests play in its formation? How much pressure has the sustainability agenda exerted? Conversely, study of this specific policy area can contribute to the exploration of the processes currently at work in development plan preparation. A detailed investigation of the factors that have brought forward and shaped HVDT-related policy can throw light on the influences likely to be at work in plan-making more generally: For instance, how important are objectors in determining policy? What aspects of sustainability are being advanced most successfully? It is with this larger context of the current nature of development plan preparation in mind, along with the HVDT planning framework discussed in Chapter 2, that a methodology for the analysis of HVDT-related policy is now developed.

CHAPTER 4

An Approach to the Analysis of HVDT-related Policy

4.1 INTRODUCTION

This book investigates HVDT-related policy as expressed in local authority development plans, by exploring the content and distribution of policy concerns throughout England and Wales and the influences at work in policy formation. The treatment of this policy area in development plans is likely to be dominated by environmental issues, which, as Chapter 2 has shown, are significant in relation to HVDT infrastructure and nearby development and are highlighted by various legislative requirements and industry-based guidelines and initiatives. Chapter 2 also indicated the possibility of tension between local planning authorities (LPAs) and the electricity industry over the environmental implications of high-voltage distribution and transmission (HVDT) activity. These associated themes, of environmental sensitivity and potential struggle between LPAs and the electricity industry, are particularly pertinent given the recent trends in development plan preparation as a whole, as discussed in Chapter 3. Here, the resurgence of plan-making in the 1990s was linked to the furtherance of sustainability objectives, encouraged partly by greater public involvement in the preparation of plans.

These themes are taken into account in this chapter, which describes the methodological approach used in this exploration of HVDT-related policy. Firstly, an overall framework for the empirical study of HVDT-related policy is developed, based upon an identification of the elements of study involved in the research, and the means by which they can be investigated. Secondly, the methods by which the study was carried out are discussed and explained.

4.2 A FRAMEWORK FOR THE EMPIRICAL STUDY OF HVDT-RELATED POLICY

4.2.1 Elements of study

This investigation of HVDT-related policy implies an interest in the following aspects of policy formulation (as implied in Figure 1.2):

- the wording and the meaning of HVDT-related policy;
- the formal procedures by which it comes to expression;
- the actors involved in policy formation, and their perspectives;
- the external features that shape actor perspectives.

These aspects closely reflect those portrayed in the representation of policy formation as shown in Figure 1.2. This figure also emphasises the possible interconnection of these elements and suggests that they fit together into an overall process of policy formation. The empirical study of HVDT-related policy therefore focuses on the process by which external features influence various actors, who then bring their perspectives to bear within the context of plan-making procedures, which give rise to specific policy wording. This understanding approximates the "institutionalist approach" to the analysis of planning practice, as advocated by Healey:

> The emerging institutionalist approach… seeks to identify both general patterns in the form and contents of planning practices and the forces that drive them. At the same time, it recognises that actual practices are deeply influenced by local histories and geographies, by the specific way in which broader forces impact on particular places, and by the capabilities of the actors involved (Healey 1997a, p. 22).

Integral to the analysis of policy formation is the administrative unit within which it is carried out. Plans are generally prepared by individual LPAs (although some structure plans are prepared jointly by neighbouring authorities); LPAs therefore provide the setting within which HVDT-related policy can be located.

However, plan-making is a process of considerable duration and complexity (Section 3.5), such that it has a significant time element. This is exemplified by the fact that some plans have been completed by LPAs different from those that began them; this has occurred in a number of cases where local government reorganisation has taken place during the making of a plan. HVDT-related policy should be seen, therefore, in the context of the history of plans as they are developed and not just as an end product of LPAs. This suggests that a more helpful unit than the LPA, in studying policy, is the "development plan process" (Hales 2000): the process by which a particular plan is initiated, progressed and brought to completion by the relevant plan-making authority.

In much of what follows, therefore, the focus of attention is upon development plan processes (DPPs) rather than LPAs. Geographically, DPPs can be equated to the LPAs responsible for them, but referring to DPPs places greater emphasis on the fact that policy is likely to have been shaped and changed throughout the protracted procedures of plan-making.

4.2.2 Levels of study

This study seeks to investigate the emergence of HVDT-related policy across England and Wales as a whole. However, it also aims to explore the precise influences at work in policy formation, with the objectives of examining the effects of geographical, social and other factors and the role of different actors. This suggests the need for a more in-depth analysis of the process of policy formation than is possible at national level. Two distinct levels of study are therefore required as follows:

1. A study of the HVDT-related policy across England and Wales, concentrating primarily on the content of policy. This implies a quantitative investigative approach, in which DPPs throughout the area are assessed for the occurrence of relevant policy, and the main policy concerns are identified.
2. A localised study of HVDT-related policy, analysing in detail the influences shaping the HVDT-related policy. This implies a more qualitative approach, in which the components and processes contributing to policy expression in specified areas are examined in depth.

Applying more than one approach to a single area of study in this way is often referred to as *triangulation* (Robson 2002). This has a number of potential advantages, particularly where quantitative and qualitative methods are combined (Bryman 1988); those applying to the present research are as follows:

- One approach provides the basis for another. The comprehensive study of policy across England and Wales can provide the basis on which to conduct the detailed analysis, suggesting possible lines of inquiry that can be pursued by means of the more localised investigation.
- The corroboration of results. This is possible insofar as different methods of investigation examine similar aspects of the research topic. In the approaches described above, mutual confirmation of results can be gained, particularly in respect of the content of policy.
- The construction of a fuller overall picture. The generalised picture provided by the England and Wales–wide study can be given greater depth by the localised study; conversely, the particulars of localised accounts can be given wider applicability by the overall study.

The adoption of two levels of research therefore offers a complementary approach in which a fuller understanding of policy formation can be gained; it is "a strategy that adds rigour, breadth and depth to any investigation" (Denzin and Lincoln 1998). The two approaches, and the precise methods adopted, are now described in more detail.

4.3 STUDY OF HVDT-RELATED POLICY ACROSS ENGLAND AND WALES

4.3.1 Data sources

A study of HVDT-related policy across England and Wales can provide a comprehensive picture of policy by

- identifying those DPPs in which HVDT-related policy has been raised;
- studying the content of this policy and categorising the main issues raised.

This approach is also concerned with the geographical location of DPPs in which HVDT-related policy concerns have been expressed.

One primary source of data on HVDT-related policy is development plan documentation: draft and final versions of plans and associated papers, reports and other documents (Figure 3.1). These indicate directly whether HVDT-related policy has been incorporated into plan preparation. The documentation also provides the precise wording of where it has been and comments upon the policy at various stages of its preparation. An extensive collection covering the whole study area, held by National Grid (NG), was made available for the purposes of this research. NG also provided a secondary source of data, an electronic database, already containing details of HVDT-related policies that have been advanced by LPAs. This secondary source allowed the large quantity of data provided by the development plan collection to be filtered, so that only those development plans containing relevant policies were looked at in detail. However, given the potential for inaccuracy and bias that is inherent in a secondary source of this kind, an assessment of the quality of the data contained in the NG database was made prior to its use. This concluded that the database was sufficiently accurate for its use in the research, though, for some aspects of the study, the information provided was verified against original sources.

In addition, the database provided an opportunity to note the extent to which objections had been brought by NG against particular policies and whether or not the LPAs concerned had resisted those objections. This was due to records within the database of NG representations and subsequent LPA responses. On the one hand, this illustrates the electricity industry's views on HVDT-related policy; on the other hand, it indicates LPA willingness either to accommodate representations or to pursue policy regardless of opposition.

4.3.2 Data analysis

<u>Research database</u>

Data from the above sources was organised and analysed by means of a research database, which was designed using the Microsoft Access software. This enabled the following steps to be taken:

1. Separate entries were set up for all DPPs throughout England and Wales.
2. Information relating to each DPP was transferred from the data sources, with additional comments as relevant.
3. Policy concerns were categorised according to frequently occurring themes.
4. Policy concerns were coded to indicate if NG had objected to them and if LPAs had subsequently resisted those objections.
5. Notes were added about the possibility of DPPs being used as case studies.

Figure 4.1 shows an example of a fictitious form taken from the database.

Further representation of data

The data were also converted into numeric form expressing, firstly, the occurrence of different aspects of HVDT-related policy throughout England and Wales and, secondly, a measure of the extent to which policy has been opposed and defended. This was then transferred to other software packages, permitting further analysis:

- Microsoft Excel: this allowed statistical and graphical representation of policy across England and Wales;
- ArcView (a Geographic Information System), with imported boundary data from UK Borders: this allowed geographic representation of policy across England and Wales.

These packages were applied, in standard ways, to convert the database information into more readable forms. This allowed interpretation of the data to take place, including the definition of policy patterns and themes apparent throughout England and Wales. The results of these analyses are presented in Chapter 5.

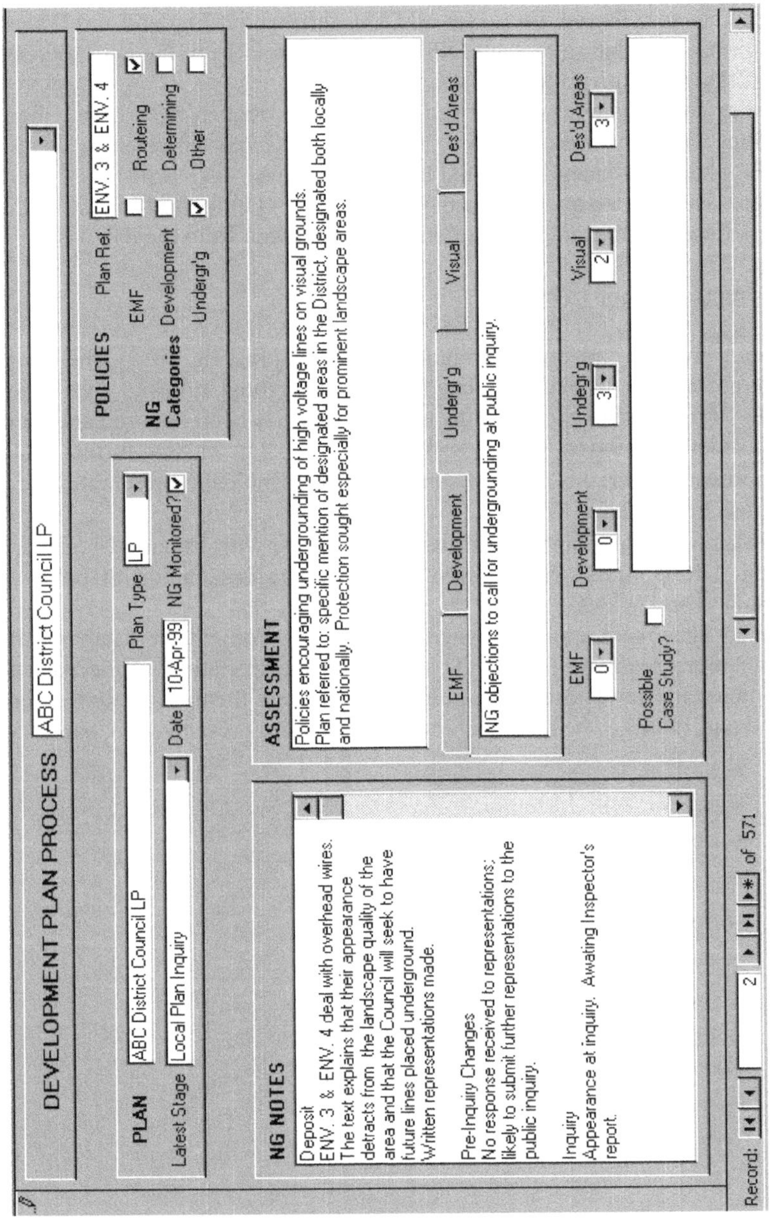

Figure 4.1 Fictitious form from the research database

4.4 LOCALISED STUDY OF HVDT-RELATED POLICY

4.4.1 The adoption of a case study approach

The second, more localised, level of study sought to investigate the content and formation of HVDT-related policy in greater detail than is possible across England and Wales as a whole. It aimed at studying the development of policy, in particular, geographical, social and other contexts, giving consideration to the involvement of different actors, and the way in which policy is shaped during plan-making procedures. This suggested the need for multiple methods of inquiry into the formation of policy in distinct settings, as is made possible by a case study approach to research.

"Case study" generally refers to a focused inquiry into one or more cases; these, in turn, can be thought of as "bounded systems", exhibiting coherence and some degree of purpose (Stake 1998). This approach usually involves the application of different methods, of a qualitative nature, in order to produce as full a description of cases as possible (Robson 2002). The local formation of HVDT-related policy is therefore well suited to case study examination. Firstly, it takes place within geographically and institutionally distinct LPAs, which provide a well-defined basis for the delimitation of cases. Secondly, policy formation involves a number of contributing elements, such as documentary material, actor involvement and environmental conditions, which require different investigative methods.

The application of a case study approach to the study of HVDT-related policy raises the question, to what extent is it possible to understand policy across the whole of England and Wales on the basis of a number of case studies? By definition, case study focuses on the particularities of a given case, but this does not preclude the possibility of generalisation from individual cases (indeed, the categorisation of policies given above assumes some commonality between DPPs). In this study, it is assumed that conclusions can be drawn, about the "population" of DPPs as a whole, that "lawful and reasonably stable relationships" between phenomena can be described on the basis of a limited number of cases (Miles and Huberman 1994, p. 8).

4.4.2 Selection of cases

Miles and Huberman advocate the study of multiple cases, in order to find underlying patterns that provide a "causal description of the forces at work" (ibid. p. 8). In addition, with regard to HVDT-related policy, it was expected that there would be more influences on policy formation than could be studied within one case. A number of case studies of DPPs was therefore chosen (the precise number being determined after the detailed analysis of DPPs throughout England and Wales (Section 5.7)).

Deciding upon the study of multiple cases leads on to the consideration of the type of case to be selected. A case study approach does not require the selected cases to be thoroughly representative of all possible cases. For example, Yin (1994) insists that generalisation from multiple cases is not of a statistical nature, as they do not have clearly defined criteria that can be statistically sampled. Rather, cases are chosen to represent the focus of research in a broader manner; this may involve choosing cases that exhibit particular features in a more pronounced form than is typical. So Patton (quoted by Flick 1998, p. 69) suggests that possible strategies for case selection include choosing those that are most interesting, critical or even extreme in some respects.

This approach was applied to the present research by selecting cases from those DPPs in which HVDT-related policy concern had been both demonstrated and pronounced. These criteria focused the attention of case study on those DPPs where there had been considerable expression of interest in HVDT-related issues and where the contributory features were likely to be most evident. The cases chosen were therefore expected to represent, in a broad sense, the forces at work in HVDT-related policy formation throughout England and Wales *where they had been intense*. In concentrating the case study on this "tier" of DPPs, it was assumed that the processes at work within them were fundamentally the same as those in DPPs where policy expression was less marked. It was expected, therefore, that principles describing policy formation drawn from the selected cases would be applicable to DPPs expressing less prominent concerns. The actual cases chosen, and the precise criteria governing their selection, are described in the next chapter (Section 5.7).

4.4.3 Methods of inquiry

As stated above, case study analysis relies upon the application of multiple methods of inquiry to produce a comprehensive description of individual cases. The precise methods used are determined by the area of study itself: the possibilities and means available within the context (Denzin and Lincoln 1998). The possible methods appropriate for the current research were determined by carrying out a pilot case study (which was later adapted to form one of the final cases); this led to the following methods of inquiry being adopted:

Documentary examination

The following documents were studied, insofar as they were relevant:
- draft and final versions of plans;
- consultation papers and reports of proposed plan changes/modifications;
- records of public representations and of evidence presented at inquiries;
- inspector/panel reports;
- other development plans, planning guidance and strategies;

- records of local planning decisions;
- historic plans.

These were examined with the following aims:

- to describe the content of HVDT-related policy and trace its development throughout the DPP;
- to identify the actors involved in policy formation and describe their perspectives;
- to place the policy in its wider policy context;
- to identify local HVDT-related planning issues.

Interview of actors

Interviews were carried out of certain actors involved in policy development, especially planning officers who were responsible for drafting and carrying forward policy. Two types of interviews were conducted.

Firstly, one key interview for each case study was carried out with the planning officer who had recently worked most closely on the policy topic. These were nearly all extended, face-to-face interviews. They were of a semi-structured nature, following an interview guide, and aimed at the following:

- to gain a fuller understanding of the LPA's HVDT-related policy concerns and the influences contributing to policy formation;
- to identify the role played by specific actors;
- to determine HVDT-related planning issues occurring locally.

Secondly, supplementary interviews were conducted, mostly by telephone. These aimed partly at seeking clarification and further detail on points already raised and also at gaining insight into other perspectives on policy. The following people were approached:

- other local authority officers;
- elected members;
- interest group representatives.

Geographical observations

In all the cases, the documentation and/or the interviews revealed localised planning issues concerning HVDT installations. These issues were explored from a geographical perspective, by means of

- a desktop map study, to note the presence of HVDT installations, and their relationship to other features, such as landscape type and urban areas;
- field observation of the LPAs' areas, to gain an impression of their geographical characteristics and the place of HVDT installations within them;
- site visits to locations where specific HVDT-related planning issues had arisen.

These observations were recorded both in the written form and by photography.

4.4.4 Case study analysis

Applying the above methods to each of the selected cases revealed detailed, and sometimes complex, processes of policy formation at work in distinct geographical settings. Separate accounts of each case study were therefore produced, before drawing together data from the cases as a whole, using a common format as given below.

- *Introduction*: description of the geographical setting, type of LPA, and so on.
- *Articulation of policy*: description of the content of HVDT-related policy and its development throughout plan preparation.
- *Key policy issues*: definition of the LPA's policy concerns and discussion of the features contributing to those concerns and the role of different actors.
- *Conclusion*: summary of the influences at work in policy formation, including a table showing the components involved in policy formation, and a network diagram showing the main links between those components.

A series of consistent "within-case" analyses was thus produced; these are presented in Chapters 6 to 8. These accounts provided the basis for a "cross-case" analysis, leading to a broader picture of policy formation (Miles and Huberman 1994).

The cross-case analysis proceeded by twin processes of data reduction and data generalisation so that common themes and processes could be described. This was carried out by:

- synthesising the components occurring throughout the cases;
- defining broad policy themes on the basis of this synthesis;
- producing generalised descriptions of those themes, using the same form of tables and network diagrams used in the within-case analyses;
- drawing comparisons between the themes.

As well as building up an overall picture of this kind, the cross-case analysis also made continued reference to the particular details of the individual cases, in order to illustrate the points being made. This analysis is described in Chapter 9.

4.4.5 "No HVDT-related policy" LPAs

The localised study of HVDT-related policy was therefore based upon the study of a number of LPAs where policy had been expressed in a pronounced form. However, this approach leaves open the question as to why policy has *not* been pursued by certain LPAs. Although investigation of these LPAs was not adopted as a primary method of inquiry, it was decided to conduct a brief survey of a number of "no HVDT-related policy" LPAs, consisting of telephone interviews of planning officers from a number of these LPAs. This survey acted as a means of testing for factors dissuading the formation of policy and provided a counterpoise to the selection of LPAs with strong expressions of policy as cases. This is described in Section 9.5.

4.5 COMBINING RESULTS

The adoption of two levels of study was based on the assumption that they would complement each other in three ways (Section 4.2.2). The first of these, that one approach would provide the basis for another, played a role in the selection of case studies (Section 4.4.2). The second, the corroboration of results, became apparent in the cross-case analysis (Section 9.2.2). The third, the construction of a fuller overall picture, relies upon the combining of results. This was carried out by drawing together the conclusions of the two different levels of study in a discursive manner, to create an overall description of the forces at work in HVDT-related policy formation throughout England and Wales. This synthesis was organised partly in accordance with the elements of policy formation suggested in Figure 1.2 (influencing features, actors, DPP and policy content) and partly along the lines of dominant policy themes that arose from the cross-case analysis. This discussion is presented in Chapter 10.

This chapter has presented a methodology for the exploration of HVDT-related policy as expressed in England and Wales, based upon two complementary scales of study. Data sources have been identified and methods of investigation and analysis that are appropriate to each level have been selected. The following chapter describes the first stage of the investigative process, the analysis of DPPs across England and Wales as a whole.

CHAPTER 5

HVDT-related Policy across England and Wales

5.1 INTRODUCTION

Chapter 4 outlined two approaches for investigating HVDT-related policy in England and Wales. This chapter describes the application of the first approach, the comprehensive study of HVDT-related policy across England and Wales.

This analysis of data relating to development plan processes (DPPs) throughout England and Wales enabled:

- the identification of DPPs that include HVDT-related policy;
- the categorisation of policy concerns within these DPPs;
- policy concerns to be related to various aspects of DPPs;
- the geographical representation of policy concerns.

The emphasis in this chapter is upon a broad understanding of HVDT-related policy: establishing the principal issues being raised, tracing geographical patterns of policy, and making connections between concerns expressed and different types and stages of plan preparation. These elements are described below; this is followed by a synthesis of these separate aspects. Finally, the results of this analysis are used as a basis for the selection of case study authorities, which are investigated in the following chapters.

5.2 DEVELOPMENT PLAN PROCESSES WITH HVDT-RELATED POLICY

5.2.1 DPPs included in the analysis

DPPs in England and Wales consist of the preparation, or review, of:

- structure plans (SPs);
- unitary development plans (UDPs);

- local plans (LPs);
- waste and mineral local plans.

An examination of data for waste and mineral local plans showed that they did not deal with HVDT issues, except for very isolated and marginal references; these plans were therefore excluded from the study. The other three types of plans were included in the analysis. A total of 433 of these types of DPP were identified throughout England and Wales; these are summarised in Table 5.1, along with the types of authority responsible for them.

However, data were not available for all of these DPPs. A number of authorities were at very early stages of plan preparation, and had not been included in the NG monitoring system at the time of the analysis. Table 5.1 shows that the number of DPPs for which data was not available amounted to only 15 (3.5%) of the total; data was therefore available for 418 (96.5%) of the DPPs considered.

5.2.2 DPPs demonstrating HVDT-related policy

The 418 DPPs for which data was available were studied to determine which of them included HVDT-related policy concerns. These DPPs were defined as containing:

- policies dealing with the planning of HVDT installations;
- policies dealing with the planning of other developments close to HVDT installations; or
- other references to the effects, or possible effects, of HVDT installations on their surroundings.

Table 5.1 Types and numbers of DPP

Type of DPP	Authorities responsible for DPP	Total number of DPPs	Number of DPPs for which data were not available	Number of DPPs for which data were available
Structure Plan	England: county authorities; some unitary and two national park authorities. Wales: two unitary authorities*.	42	1	41
Unitary Development Plan	England: London and metropolitan borough authorities; some other unitary authorities. Wales: national park and most unitary authorities.	96	5	91

Type of DPP	Authorities responsible for DPP	Total number of DPPs	Number of DPPs for which data were not available	Number of DPPs for which data were available
Local Plan	England: district, national park and some unitary authorities. Wales: two unitary authorities*.	295	9	286
Total		**433**	**15**	**418**

* At the time of the analysis, Powys County Council and Rhondda Cynon Taff County Borough Council had not progressed to UDP preparation, and so were considered under the two-tier system of SPs and LPs still in force in their areas.

Figure 5.1 shows that 158 (37.8%) of the DPPs demonstrate HVDT-related policy concern of some kind; in other words, over a third of the DPPs included in the analysis deal in some way with HVDT planning issues. Moreover, all three types of DPP express policy concern, though not in equal proportions. LPs raise HVDT-related issues more frequently (in 42.2% of cases) than do UDPs (26.6%) or SPs (24.4%).

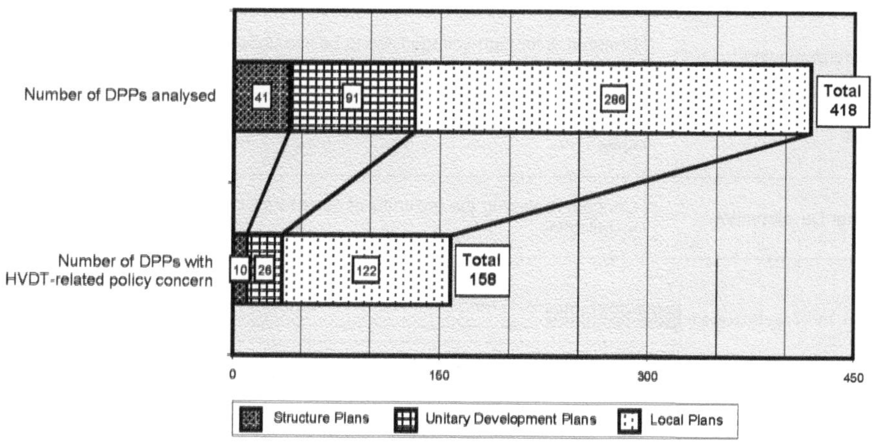

Figure 5.1 HVDT-related policy concern in DPPs

5.3 HVDT-RELATED POLICY CONCERNS

5.3.1 Occurrence of policy concerns

The 158 DPPs that demonstrated HVDT-related policy of some kind were further analysed to determine the broad concerns that each of them raised. A study of the data suggested a number of issues that occurred consistently throughout these DPPs; a reference to the wording of policies enabled generic descriptions to be drawn up for each of these issues. These are shown in Table 5.2. These concerns address aspects of the sighting either of HVDT installations, or of other development close to HVDT installations. They focus on the potential effects of HVDT installations on their immediate surroundings, and demonstrate interest in local, rather than more strategic, planning issues.

These policy concerns provide categories into which the DPPs can be organised. Figure 5.2 shows the number of DPPs falling within each category. This indicates that "visual effects" is a matter of concern in a large proportion (86.1%) of the 158 DPPs, as is "undergrounding" (70.3%). "Designated areas" is raised in about half of the DPPs in question (53.2%). About a third of the DPPs refer to "EMF" (31.6%) and "other development" (37.3%).

Table 5.2 HVDT-related policy concerns in DPPs

Abbreviation	Policy Concern
Visual Effects	Concern regarding the visual effects of HVDT installations
Designated Areas	Concern to protect designated areas from the effects of HVDT installations
Undergrounding	Preference for high-voltage lines to be laid underground rather than suspended overhead
EMF	Concern regarding the possible effects of EMF emitted from HVDT installations
Other Development	Concern regarding the proximity of other forms of development to HVDT installations

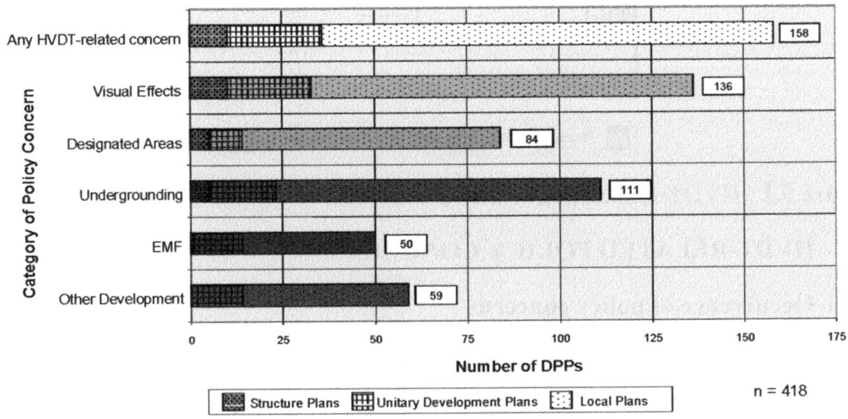

Figure 5.2 Occurrence of HVDT-related policy concern in DPPs

5.3.2 Association of policy concerns

It is evident from Figure 5.2 that many of the DPPs include more than one policy concern. This raises the possibility that different issues are associated with each other. To explore this, Table 5.3 indicates the extent to which each of the policy concerns occurs in conjunction with others. The columns of this table reveal, for any given concern, which of the other concerns it is most associated with. For example, in those DPPs where "visual effects" occurs, "designated areas" also occurs in 60% of them.

Table 5.3 Percentage association of HVDT-related policy concerns

		Percentage association with second policy concern				
		Visual Effects	Designated Areas	Under-grounding	EMF	Other Development
Second policy concern	Visual Effects		98	94	66	69
	Designated Areas	60		64	36	42
	Undergrounding	76	85		64	63
	EMF	24	21	29		63
	Other Development	30	30	33	74	

This table suggests that there are a number of identifiable associations among the five policy concerns; these are portrayed more clearly in Figure 5.3. This diagram suggests that:

- "designated areas" and "undergrounding" are very strongly associated with "visual effects", and "designated areas" is strongly associated with "undergrounding"; these three policy concerns are all associated with each other.
- "EMF" and "other development" are associated with each other; they are also both associated with "visual effects" and "undergrounding".

There are, therefore, two groups of significant association among the policy concerns. These can be seen in Figure 5.3 as:

- the top triangle of concerns;
- the bottom pair of concerns, along with its connections to "visual effects" and "undergrounding".

5.4 HVDT-RELATED POLICY AND ASPECTS OF DPPS

5.4.1 Policy concerns and type of DPP

Figure 5.2 indicates that nearly all of the five policy concerns are raised in each of the three different types of DPP: SPs, UDPs and LPs. However, they are not all raised in each type of plan in equal proportions. In order to explore this, Figure 5.4 shows the relative occurrence of each of the policy concerns in each type of DPP. This indicates that:

- "visual effects" is raised in similar percentages of all three types of plan;
- "designated areas" is raised in similar percentages in SPs and LPs, but slightly less in UDPs;

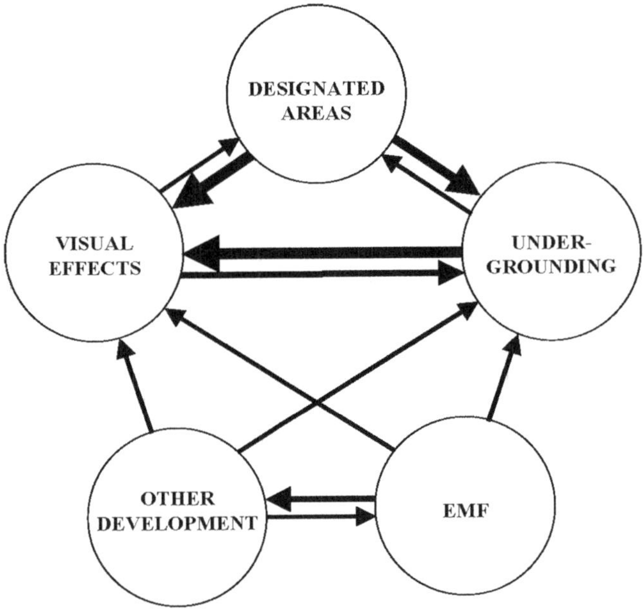

Figure 5.3 Strengths of association between HVDT-related policy concerns

An arrow pointing from one concern to another indicates that a high percentage (60% or more) of the first concern's occurrences in DPPs coincides with occurrences of the other concern. The thickness of the arrows represents the percentage values given in Table 5.3

- "undergrounding" is raised in similar percentages in UDPs and LPs, but slightly less in SPs;
- "EMF" and "other development" are raised very little in SPs; they are raised in a greater percentage of UDPs than LPs.

5.4.2 Objections to, and defence of, policy concerns

As well as giving information on the content of HVDT-related policy in DPPs across England and Wales, the data also provided a record of objections that NG had made to proposed policies, and the response of LPAs to those objections. This provided an indication of both the reaction of the electricity industry to the issues being raised by LPAs, and of the extent to which LPAs were prepared to defend their concerns against formal objections. The DPPs that included HVDT-related policy were therefore analysed to determine:

- whether or not NG had brought formal objections against those expressions of policy;
- where such objections had been brought, whether or not the LPA had resisted those objections and defended their policy concerns.

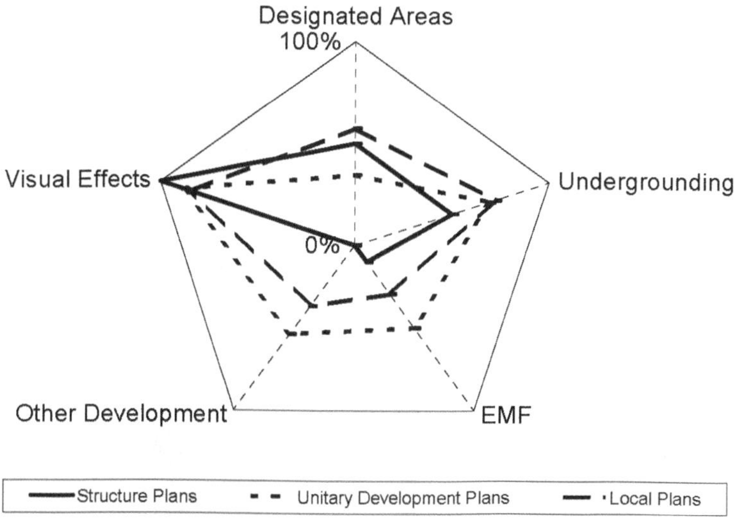

Figure 5.4 Occurrence of HVDT-related policy concerns in different types of DPP

The possible occurrence of these actions within the DPPs provides a measure of the degree of conflict that has been provoked by LPAs seeking to advance HVDT-related policy.

NG objections were directed typically against particular aspects of policy. These coincided to a large extent with the policy concerns identified above, with the exception of "visual effects", which could not easily be linked to objections. The other policy concerns were, however, clearly identifiable both in NG objections and LPA policy defence. Figure 5.5 shows, for each of these four concerns, the relative proportions of DPPs where objections were made and where they were resisted. The defence of policy by LPAs was taken to be particularly significant at the quasi-judicial stage of plan preparation (local public inquiry in the case of UDPs and LPs, and examination in public for SPs).

Figure 5.5 shows that:

- "Designated areas" is the policy concern that provoked the lowest proportion of objections from NG, in less than half of the DPPs where it occurred.
- The other three policy concerns provoked objections from NG in at least three-quarters of the DPPs where they occurred; the greatest proportion of objections were made in relation to "EMF".
- LPAs resisted NG's objections in about a third of the cases where "designated areas" and "undergrounding" were concerned, and in about half of the cases where "EMF" and "other development" were concerned.

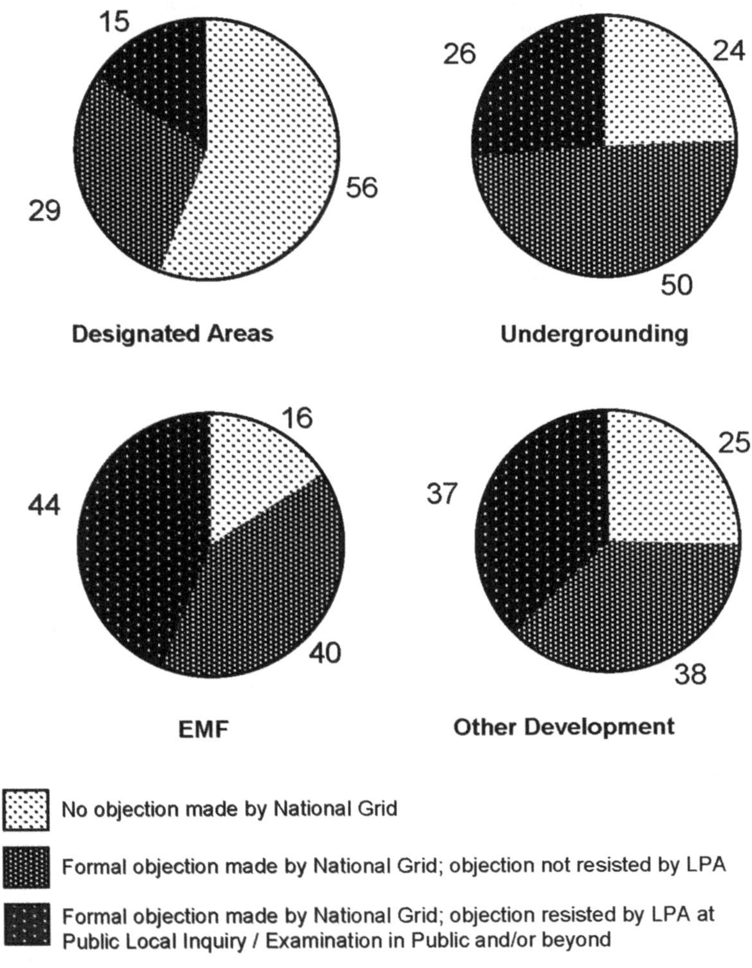

Figure 5.5 Objections to, and defence of, HVDT-related policy concerns

5.5 GEOGRAPHICAL DISTRIBUTION OF HVDT-RELATED POLICY

In order to investigate the geographical distribution of HVDT-related concerns, the data presented above were also represented spatially, using a geographical information system (GIS). This enabled the occurrence of HVDT-related policy across England and Wales to be shown, by:

- policy concern;
- type of DPP;
- extent of NG objections and LPA defence.

Figure 5.6 portrays the demonstration of policy in SPs. (Only four policy concerns are shown here, because "other development", was not raised in SPs; see Figure 5.2.) Figures 5.7–5.11 portray the demonstration of all five policy concerns in UDPs and LPs. The maps also indicate, where relevant, the DPPs where formal objections were brought against proposed policy by NG, and where LPAs resisted those objections.

5.5.1 Summary of the geographical distribution of policy

The distribution of DPPs demonstrating HVDT-related policy as shown in Figures 5.7–5.11 can be described with reference to the following criteria:

- The type of distribution pattern: clusters (small, medium or large) or dispersed.
- The areas where policy concerns mainly occur.
- The areas where NG objections and LPA policy defence are most pronounced.

Table 5.4 contains the description of the maps according to these criteria. The following main points can be inferred from this table:

<u>Visual effects</u>

This policy concern occurs strongly in SPs in northeast and eastern England; moreover, these processes show a marked degree of clustering. UDPs and LPs also demonstrate this concern mainly in clusters, the largest and strongest of these being in parts of northern, eastern, southwest and southern England and of Wales. Particularly extensive clusters are centred on northwest England and East Anglia.

<u>Designated areas</u>

A cluster of SPs in northeast England demonstrates this policy concern. UDPs and LPs demonstrate this concern mostly in clusters, especially in parts of northern, eastern and southwest England; however, there is some additional dispersal of concern. The pattern corresponds partly to that for "visual effects". NG objections and LPA policy defences are most pronounced in parts of eastern England, especially North Yorkshire and East Anglia.

Undergrounding

This policy concern occurs in a dispersed pattern in SPs, though this includes a small cluster in northeast England. The pattern of occurrence among UDPs and LPs is similar to that of "visual effects", though with slightly less extensive clusters; these are most marked in parts of northern, eastern, southwest and southern England and of south Wales. NG objections and LPA policy defences show a dispersed pattern, though are most pronounced in parts of northern England, especially North Yorkshire.

EMF

Only one SP demonstrates this policy concern. UDPs and LPs demonstrate this concern in a dispersed pattern, though this includes a number of small clusters, especially in North Yorkshire, South Yorkshire, Buckinghamshire-Cambridgeshire and around the Thames Estuary. NG objections and LPA policy defences are most pronounced in these clusters, though are also found in other dispersed occurrences of the policy concern.

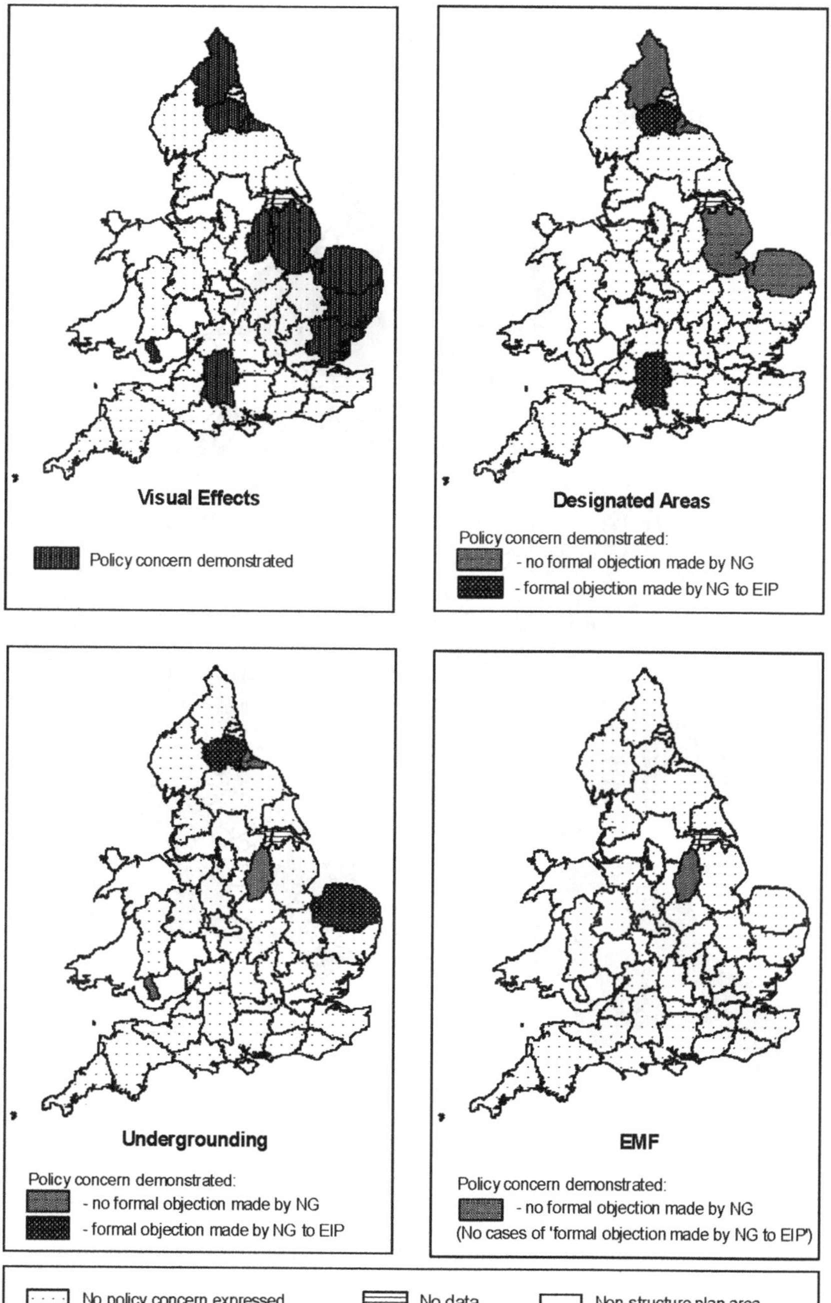

Figure 5.6 HVDT-related policy concerns in structure plan areas

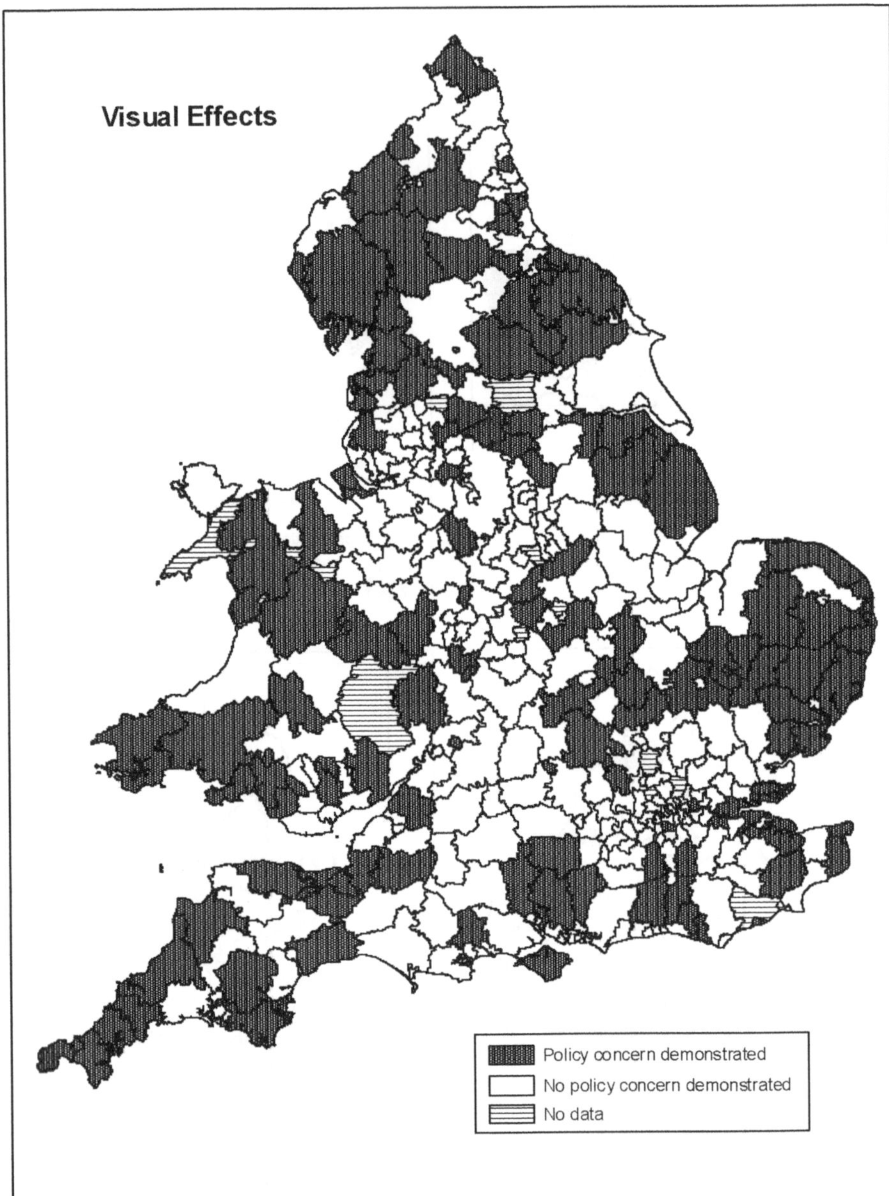

Figure 5.7 Policy concern regarding "visual effects" in UDP and LP areas

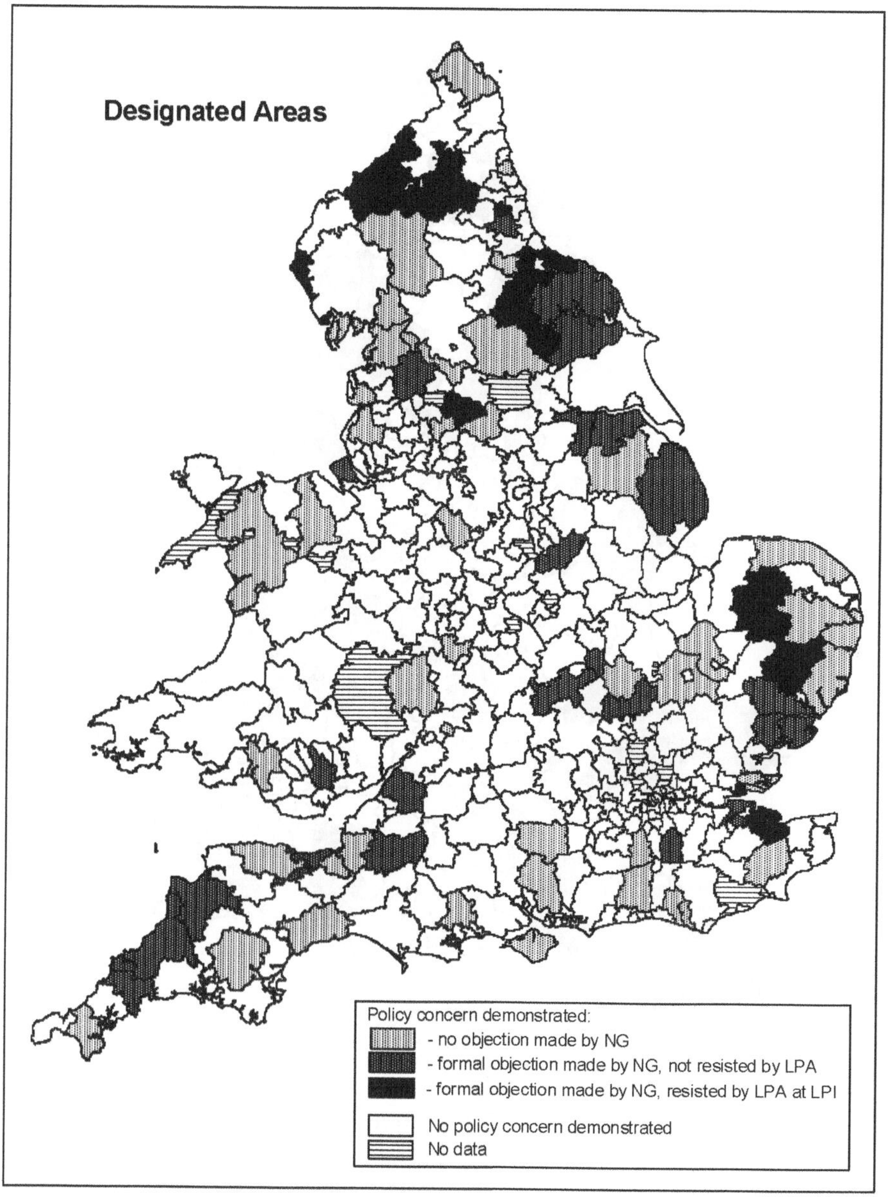

Figure 5.8 Policy concern regarding "designated areas" in UDP and LP areas

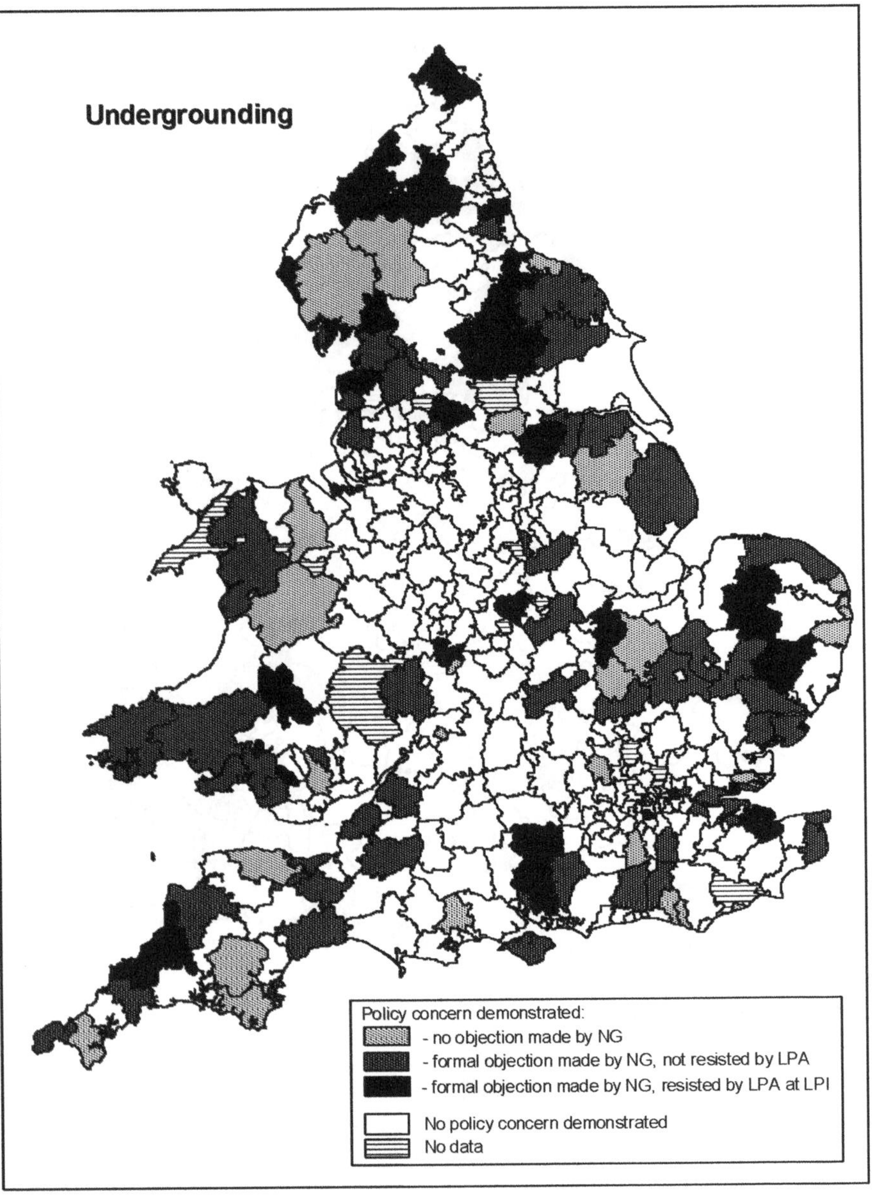

Figure 5.9 Policy concern regarding "undergrounding" in UDP and LP areas

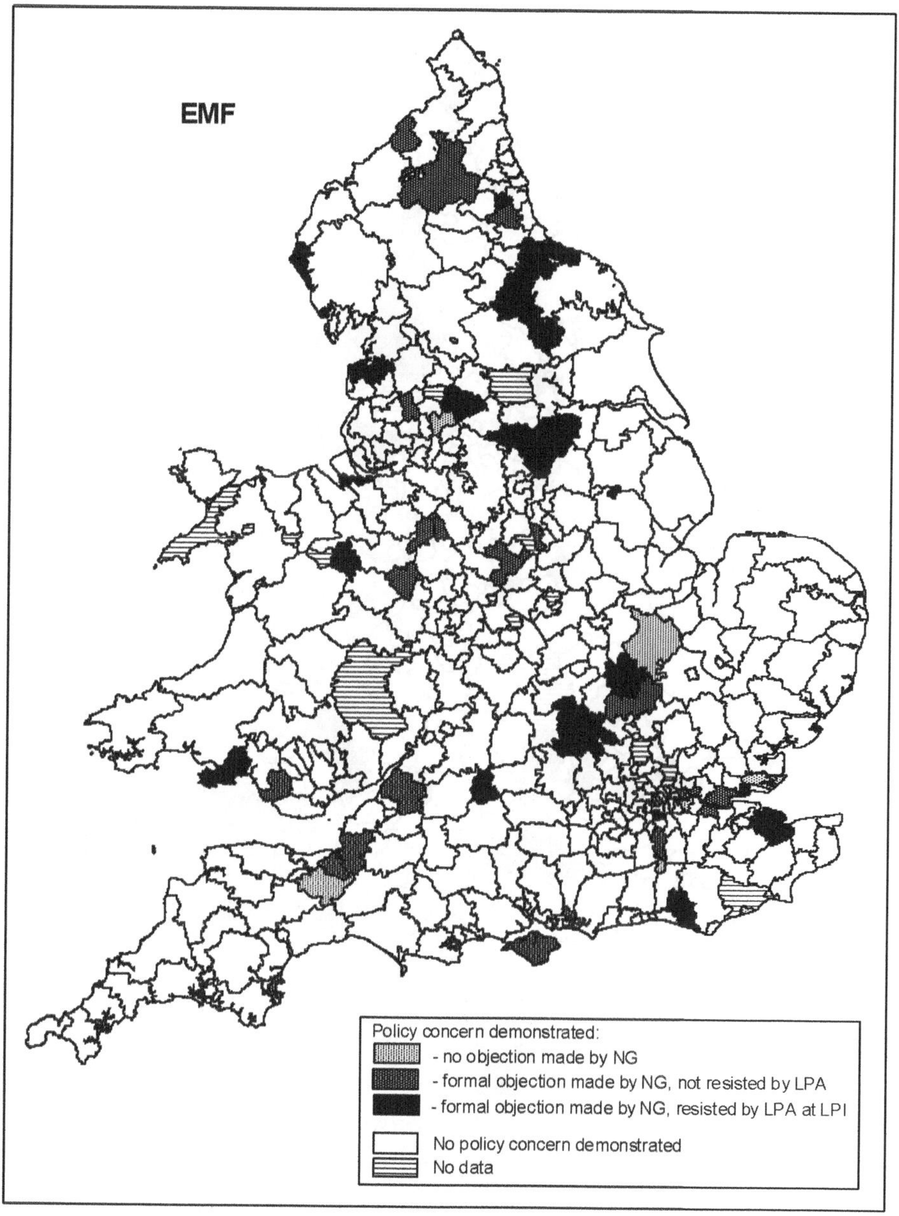

Figure 5.10 Policy concern regarding "EMF" in UDP and LP areas

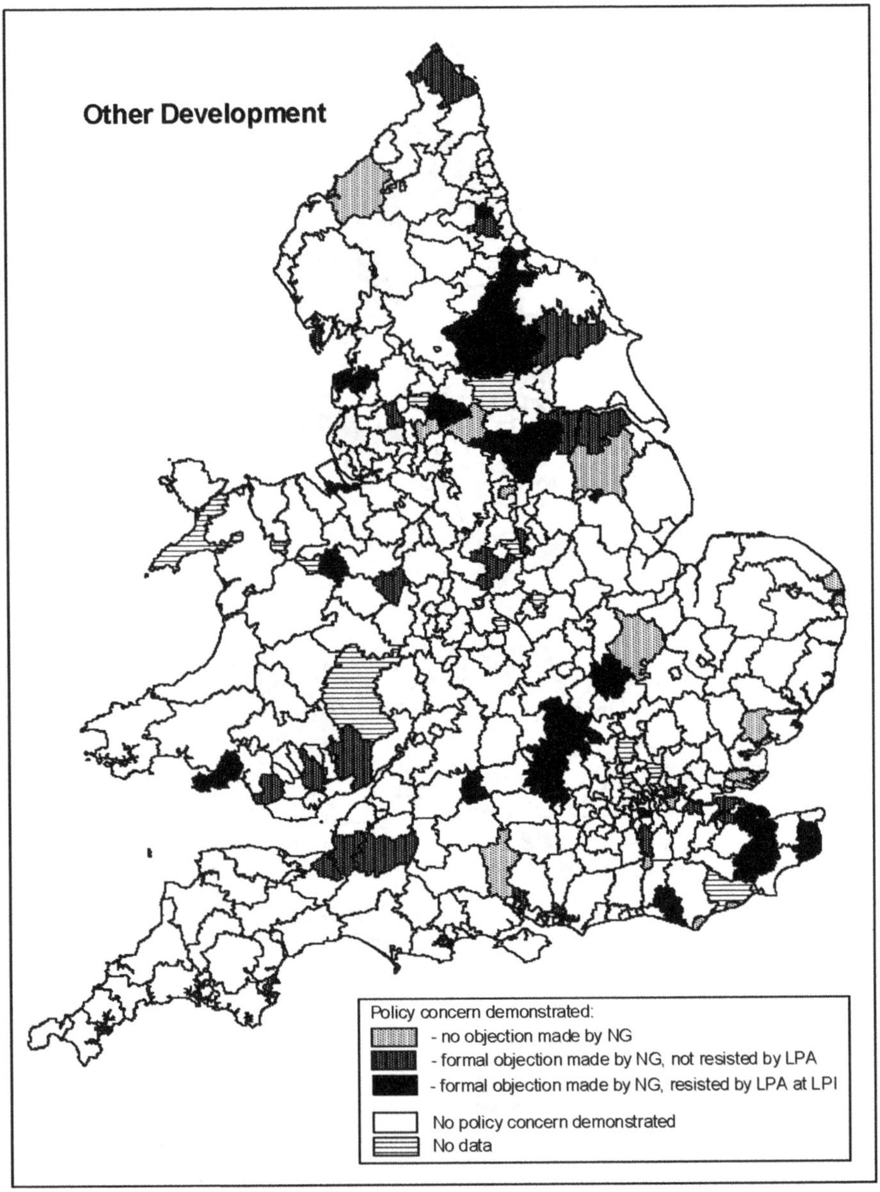

Figure 5.11 Policy concern regarding "other development" in UDP and LP areas

Table 5.4 Spatial distribution of HVDT-related policy concerns

Policy concern	Type of DPP	Type of distribution pattern	Areas where policy concerns mainly occur*	Areas where NG objections and LPA policy defence are most pronounced
Visual Effects	**SPs** (Fig. 5.6)	Clustered and dispersed	• Clusters in northeast England, Nottinghams hire–Lincolnshire and East Anglia. • Dispersed occurrence elsewhere.	Not applicable.
	LPs & UDPs (Fig. 5.7)	Mostly large and medium-sized clusters	• Northern England: clusters merging, centred on Lancashire-Cumbria, North Yorkshire, South Yorkshire and north Lincolnshire. • Wales: clusters in north and south Wales. • Eastern England: clusters merging across the East Midlands and East Anglia. • Southern England: clusters on the southwest peninsula, along the south coast and around the Thames Estuary.	Not applicable.
Designated Areas	**SPs** (Fig. 5.6)	Clustered and dispersed	• Cluster in northeast England. • Dispersed occurrence elsewhere.	• County Durham and Wiltshire.
	LPs & UDPs (Fig. 5.8)	Medium-sized clusters and some dispersal	• Northern England: clusters centred on Lanca-shire-Cumbria, North Yorkshire, South Yorkshire and north Lincolnshire. • Eastern England: clusters centred on Bedford-shire and in East Anglia. • Southern England: clusters on the southwest peninsula, along the south coast and around the Thames Estuary. • Some dispersed occurrences in Wales and the Midlands.	• Cumbria-Northumberland, North Yorkshire, East Anglia and the Thames estuary.

Policy concern	Type of DPP	Type of distribution pattern	Areas where policy concerns mainly occur*	Areas where NG objections and LPA policy defence are most pronounced
Undergrounding	**SPs** (Fig. 5.6)	Clustered and dispersed	• Cluster in northeast England. • Dispersed occurrence elsewhere.	• County Durham and Norfolk.
	LPs & UDPs (Fig. 5.9)	Mostly large and medium-sized clusters	• Northern England: clusters merging, centred on Lancashire-Cumbria, North Yorkshire and north Lincolnshire. • Wales: clusters in north and south Wales. • Eastern England: clusters merging across the East Midlands and East Anglia. • Southern England: clusters on the southwest peninsula, along the south coast and around the Thames Estuary.	• Clusters in Lancashire, Cumbria-Northumberland, North Yorkshire, East Anglia, Hampshire. • More dispersed objections and defence elsewhere.
EMF	**SPs** (Fig. 5.6)	One occurrence	• Nottinghamshire.	Not applicable.
	LPs & UDPs (Fig. 5.10)	Dispersed, with small clusters	• Small clusters in North Yorkshire, South Yorkshire, Buckinghamshire-Cambridgeshire, Thames Estuary, Somerset. • Dispersed occurrence elsewhere.	• Small clusters in North Yorkshire, south Yorkshire, Buckinghamshire-Cambridgeshire, Thames Estuary. • Dispersed objections and defence elsewhere.

Policy concern	Type of DPP	Type of distribution pattern	Areas where policy concerns mainly occur*	Areas where NG objections and LPA policy defence are most pronounced
Other Development	**SPs** (Fig. 5.6)	None	Not applicable.	Not applicable.
	LPs & UDPs (Fig. 5.11)	Dispersed, with small / medium clusters	• Small / medium clusters in North Yorkshire, West Yorkshire – north Lincolnshire, Thames Estuary, Somerset. • Dispersed occurrence elsewhere.	• Clusters in North Yorkshire, south Yorkshire, Buckinghamshire-Cambridgeshire, Thames Estuary. • Dispersed objections and defence elsewhere.

Other development

No SP demonstrates this policy concern. UDPs and LPs demonstrate this concern in a fairly dispersed pattern, though with a number of small and medium-sized clusters. These are in similar locations to those identified for "EMF", though the Yorkshire clusters are larger. Also, as for "EMF", NG objections and LPA policy defences are most pronounced in the clusters, although they are also to be found in other dispersed locations.

5.6 PATTERNS IN HVDT-RELATED POLICY

The above results have been presented, firstly, in a sequential manner, showing the:

- identification of DPPs with HVDT-related policy;
- definition of policy concerns;
- association of different policy concerns;
- differences relating to DPP type;
- extent of objections and policy defence.

Secondly, these same results have been portrayed and described geographically.

These results suggest various patterns of policy content and distribution. This section aims to explore those patterns further by drawing upon the results as a whole, particularly where similar themes are emerging from the different stages of data analysis. This allows a more unified picture to be drawn of HVDT-related policy across England and Wales, providing a basis for the ensuing case studies.

5.6.1 Spread of policy

One hundred and fifty eight DPPs demonstrate HVDT-related concern (5.2.2). Figures 5.6-11 show these to be spread widely throughout England and Wales. However, the distribution of these DPPs is not even, or random; there are strong degrees of clustering in certain regions, particularly for certain policy concerns. The following summary can be made about the overall spread of policy:

- HVDT-related concerns are not universal, but they are widely expressed throughout England and Wales.
- HVDT-related concerns are most frequent in certain areas or regions, often in groups of neighbouring authorities.

5.6.2 Variations in the expression of policy concerns

There are five distinct areas of policy concern, which have been defined on the basis of policy wording (Table 5.2). These concerns show variations in the ways that they are expressed in the DPPs, in respect of:

- their occurrence in the DPPs (Figure 5.2);
- their occurrence in different types of DPP (Figure 5.4);

- the objections that they provoke from NG and the defence they receive (Figure 5.5);
- their patterns of geographical distribution (Table 5.4).

These variations are summarised in Table 5.5.

Table 5.5 Characteristics of policy concerns as expressed in DPPs

	Relative level of occurrence in DPPs				Relative level of objections/defence		Geographical distribution
	All types	LPs	UDPs	SPs	NG objections	LPA defence	
Visual Effects	High	High	High	High	NA	NA	Large/medium clusters
Designated Areas	Medium	Medium	Low	Medium	Low	Low	Medium clusters and dispersal
Under-grounding	High	High	High	Medium	Medium	Medium	Large/medium clusters
EMF	Low	Medium	High	Low	High	High	Dispersed and small clusters
Other Development	Low	Medium	High	Low	Medium	High	Dispersed and small/medium clusters

1. This table shows that "visual effects" and "undergrounding" share similar levels of occurrence and have similar geographical distributions. They have high levels of occurrence in DPPs as a whole, across the range of types of DPP. They have the most widespread pattern of geographical distribution, forming large clusters; these predominantly cover rural areas, but also some urban areas. This implies that these concerns are focused mostly, but not exclusively, on the effects of HVDT installations in rural areas; it also suggests that undergrounding lines are seen as a means of offsetting their visual effects. "Undergrounding" provokes a medium level of contestation between NG and LPAs.
2. "Designated areas" have a medium level of occurrence in DPPs as a whole, occurring more in SPs and LPs, which predominantly cover rural areas. The rural emphasis of this concern is also evident in its geographical distribution, which is widely spread and clustered, but not as extensive as the distributions of "visual effects" and "undergrounding". The definition of this concern also suggests that it is likely to be have a predominantly countryside focus, given the location of the most extensive designated areas. "Designated areas" provokes a relatively low level of contestation between NG and LPAs.
3. "EMF" and "other development" share similar levels of occurrence and have comparable geographical distributions. They have the lowest level of occurrence in DPPs as a whole, but occur relatively frequently in UDPs, the majority of which cover metropolitan areas, and rarely in SPs, which cover more rural county areas. Their more dispersed pattern of distribution reflects this tendency to occur in smaller, more urban, areas. This is consistent with the definitions of these two concerns, which emphasise the possible effects of

HVDT installations on nearby development and people. "EMF" and "other development" both provoke high relative degrees of contestation between NG and LPAs.

5.6.3 Grouping of policy concerns

Two groups of association are evident among the five policy concerns (Figure 5.3), suggesting two distinct, though interconnected, policy areas, dealing with the effects of HVDT installations in countryside and urban areas respectively.

1. "Designated areas" have a strong countryside focus. "Visual effects" and "undergrounding" are strongly associated with it and each other; they also share the geographical distribution of "designated areas" (although they extend further). There is a body of policy with overlapping concerns for the visual effects of HVDT installations in rural areas, a sense of protection for designated areas, and a preference for undergrounding. This policy area provokes some contestation between NG and LPAs, especially with regard to "undergrounding".
2. "EMF" and "other development" have an urban focus. As well as being associated with each other, they are also associated with "visual effects" and "undergrounding", insofar as these extend into built-up areas. There is a body of policy with overlapping concerns for the visual and possible EMF effects of HVDT installations in urban areas, unease regarding the proximity of development to HVDT installations, and a preference for undergrounding. This policy area provokes a relatively high degree of contestation between NG and LPAs, especially with regard to "EMF" and "other development".

5.7 CASE STUDY SELECTION

Chapter 4 set out the theoretical basis for the second approach to examining the formation of HVDT–related policy, that of case study investigation (Section 4.4.1). According to this, the selection of case studies is dependent upon knowledge gained about potential cases from the overall study of HVDT-related policy across England and Wales; the above analysis provides this basis for the choice of cases.

The following categories have emerged as relevant to HVDT-related policy:

- occurrence of HVDT-related policy;
- particular policy concerns and broad policy areas;
- type of DPP;
- objections to and defence of policy concerns;
- geographical distribution of policy concerns.

Cases were therefore sought that reflected as full a range of these categories as possible. In addition, a number of possible factors affecting policy formation suggested themselves during the course of analysing the data; these included the possible influence of a major transmission line project, the active role of interest groups, and broader environmental strategies to plan-making. The selection of cases also took into account these particularities of individual DPPs. Finally, cases were favoured that involved stronger expressions of policy formation, where the factors involved are likely to be more pronounced (Section 4.4.2).

The number of case studies selected was determined, on the one hand, by the need to provide an adequate coverage of the above categories and factors, and, on the other hand, by the need to conduct a sufficiently in-depth study of each case chosen. Miles and Huberman present a study involving 12 cases as an example of one that enables sufficient generalisation (1994, p. 31). For the current research, 10 cases were reckoned to be sufficient to cover the range of issues to be explored, yet not too many to prevent detailed investigation of each case.

The final choice of case studies is shown in Table 5.6 and in Figure 5.12. This shows that attention was also given to achieving a reasonable geographical spread across England and Wales. Some attention was also given to the settlement pattern of the LPAs' areas, following the identification of separate policy areas relating to countryside and urban concerns. This allowed the cases to be arranged into three groups:

- cases 1-3: urbanised areas;
- cases 4-6: rural areas;
- cases 7-10: town-rural areas.

Although this categorisation is not a primary means of explaining policy, it does reflect some features occurring in common between cases, and provides a convenient means of organising the case study accounts. The individual case studies are therefore presented in these three categories in Chapters 6-8.

5.8 CONCLUSION

This chapter has provided an outline of HVDT-related policy as found across England and Wales. Over a third of LPAs are advancing some aspect of this policy area through the preparation of development plans. They are focusing their concern on a number of issues that are of immediate, local interest and relevance; some of these are more common than others. Policy expression is widespread across England and Wales; however, it is particularly concentrated in certain areas, with groups of neighbouring LPAs expressing similar concerns. There is also a distinction in the issues that are being addressed in urban and rural areas; these two settings give rise to different emphases of policy. HVDT-related policy is giving rise to some contestation between LPAs and the electricity industry, especially in relation to certain policy concerns.

As well as presenting an overall picture of HVDT-related policy across England and Wales, this analysis also provided the empirical basis for the selection of case studies. These were chosen to represent the categories that were found to be relevant to HVDT-related policy:policy content, geographical distribution, and so on. The themes revealed in this analysis of policy across England and Wales therefore provide a background for the more detailed investigation of policy formation that the case studies seek to carry out in the following chapters.

The results presented in this chapter are also drawn upon following the case study analysis, as individual elements of policy formation are discussed in greater depth in Chapter 10.

Table 5.6 Case study selection

	Local Planning Authority	DPP Type	Predominant policy concerns Strong LPA defence (underlined)	Geographical distribution of predominant policy concerns	Possible factors affecting policy formation	Settlement pattern
1	**Swindon Borough Council**	LP	"<u>EMF</u>", "<u>Other development</u>"	Southern England dispersed occurrence	Interest group activism	Predominantly urban
2	**Rotherham Metropolitan Borough Council**	UDP	"<u>EMF</u>", "<u>Other development</u>"	South Yorkshire cluster	Policy concerns spreading between authorities; high density of HVDT installations	Urban
3	**Newham Council**	UDP	"Visual", "<u>Undergrounding</u>", "EMF", "Other development"	Thames estuary cluster	Brownfield regeneration affected by HVDT installations	Urban
4	**Tynedale Council**	LP	"Visual", "<u>Designated areas</u>", "<u>Undergrounding</u>"	Northern England cluster	Extensive designated areas and high-quality landscape	Rural
5	**Norfolk County Council**	SP	"Visual", "Designated areas", "Undergrounding"	Eastern England cluster/dispersed occurrence	Extensive rural and designated areas	Predominantly rural
6	**Snowdonia National Park Authority**	LP	"Visual", "Designated areas", "Undergrounding"	North Wales cluster/dispersed occurrence	National park status	Rural
7	**Redcar and Cleveland Borough Council**	LP	"Visual", "<u>Designated areas</u>", "Undergrounding", "<u>EMF</u>"	North Yorkshire cluster	Major, controversial HVDT project	Urban-Rural
8	**Mendip District Council**	LP	"Visual", "Designated areas", "Undergrounding", "Other development"	Southwest England cluster	Strategic environmental approach to plan-making	Town-Rural
9	**Aylesbury Vale District Council**	LP	"Visual", "<u>EMF</u>", "<u>Other development</u>"	Bedfordshire – Cambridgeshire cluster	Development pressure near HVDT installations	Town-Rural
10	**Bridgend County Borough Council**	UDP	"Visual", "Undergrounding", "EMF", "Other development"	South Wales cluster/dispersed occurrence	History and wider framework of HVDT-related policy	Town-Rural

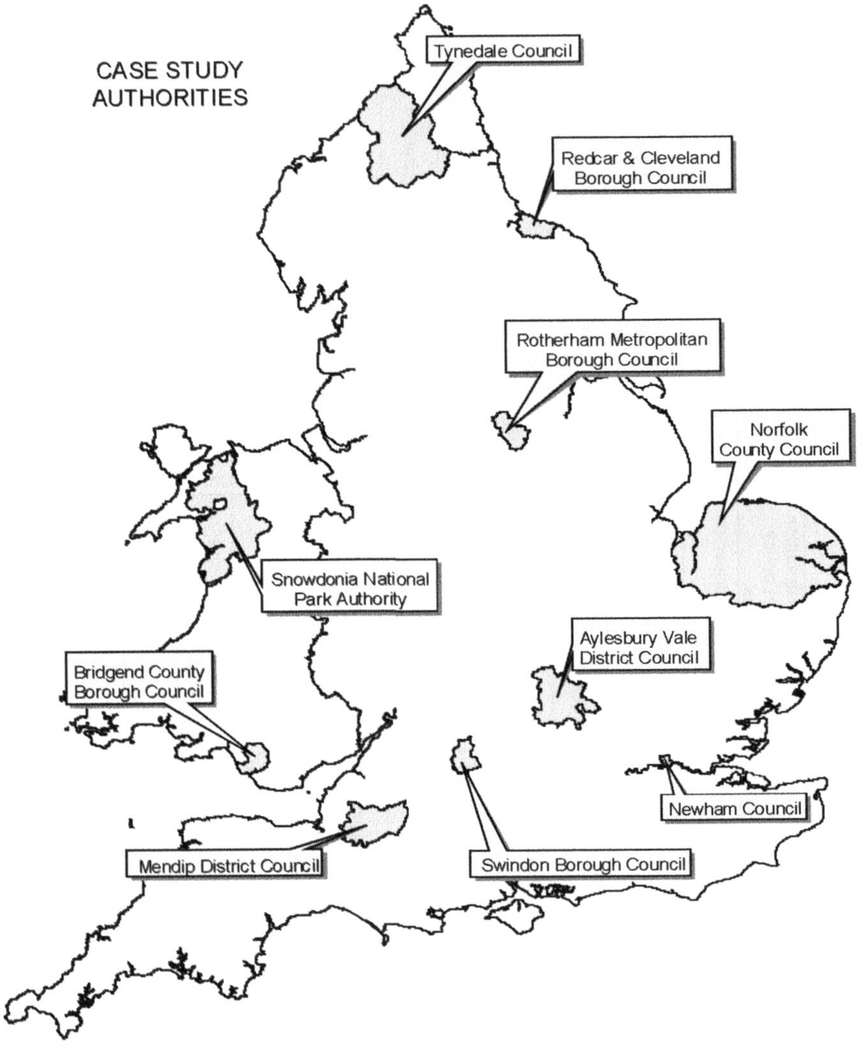

Figure 5.12 Case study authorities

CHAPTER 6

Case Studies (1): Urbanised Areas

6.1 INTRODUCTION

This chapter and the following two chapters give accounts of the 10 case studies that form the basis of the in-depth exploration of HVDT-related policy (Section 4.4.1); they are organised into these three chapters according to their overall settlement patterns. The descriptions of the case studies follow a common format. Extensive reference is made throughout these accounts to interviews with actors involved in policy formation. An analysis of the case studies as a whole follows in Chapter 9.

The case studies in this chapter explore the concerns of local planning authorities (LPAs) in heavily urbanised settings. Swindon Borough Council's (SBC)'s area includes the large town of Swindon; although there is also an extensive rural tract in the District, the council's HVDT-related policy relates solely to the urban area. Rotherham Metropolitan Borough Council's area is dominated by the town of Rotherham and adjoining settlements. Newham Council covers a densely populated part of London. The three authorities concentrate primarily on the presence of existing HVDT installations in their areas and on planning difficulties arising from their proximity to other development, both existing and proposed; different solutions are proposed to overcome the perceived difficulties. These case studies are characterised by a relatively high degree of public and elected member involvement in policy-making, and by some contestation with the electricity industry.

SWINDON BOROUGH COUNCIL

6.2 INTRODUCTION

Swindon Borough Council is a district authority in Wiltshire in southwest England. The district is dominated by the town of Swindon, which is centrally located in the area and surrounded by a rural hinterland.

High Voltage Electricity Installations S. Jay

Throughout its recent history, the town of Swindon has experienced periods of rapid growth and a changing economic base, associated with significant population increases. Over the last few decades, there have been repeated drives for economic diversification and urban expansion, partly in response to a national aim to relieve pressure in London and more lately assisted by the development of the M4 and high-speed rail links. Consequently, the population of the town has nearly doubled since 1950 to over 170,000; moreover, a further major expansion is planned, in line with structure plan allocations for housing and employment.

SBC adopted its first district-wide local plan in 1999. The plan makes provision for the intended expansion of Swindon, mainly through the development of a single, large-scale greenfield site, and also speaks of the "enormous strain on the area's capacity to cope with and assimilate further change" (SBC 1999). The plan therefore lays stress on the environmental aspects of new development, asserting that growth must increasingly be measured not just in terms of quantity but also of quality and diversity. This case study was carried out during the early stages of review of the plan.

6.3 ARTICULATION OF POLICY

SBC has brought a policy relating overhead lines (OHLs) to adoption within the "Housing" section of its plan (SBC 1999). The policy is concerned with the proximity of new housing to existing 132 kV OHLs and sets a corridor along the length of lines within which residential development should not be built. The policy is justified primarily on the grounds of amenity considerations, but reference is also made to the possible effects of EMF upon human health.

This policy found its first formal expression in Supplementary Planning Guidance adopted by the council (then known as Thamesdown Borough Council (TBC)) in 1992. This guidance established an "exclusion zone" along the length of 132 kV OHLs within which new dwellings and their curtilages would not be permitted. The width of the zone was set at a minimum of 15 m, thus defining the area lying broadly beneath the conductors; the guidance says that it is "environmentally unacceptable" to allow residential development within this zone. Furthermore, the guidance states that, "in the interests of amenity and safety", housing would not be permitted within a further 5 m of the exclusion zone or of pylons; this distance was calculated to take into account the possible swing of conductors from their centre line in adverse weather conditions (TBC 1992).

At the time of adopting this supplementary guidance, the council was also at the early stages of preparing its local plan (TBC 1994). At this point, however, the plan did not deal with the issue of development near OHLs. The later inclusion of the development constraint policy came about following representations on the deposit draft of the plan made by Swindon Friends of the Earth (FoE); they objected to the omission of the supplementary guidance from the plan. They also argued that the guidance, in the form of a plan policy, should make explicit reference to the EMF effects of OHLs. Finally, they called for the exclusion zone to be increased to a width of at least 50 m wherever "high readings" of EMF are obtained from an OHL, as a precautionary approach.

In response to these representations, the council proposed the insertion of a new policy into the local plan. This was based on the same arguments of amenity and safety as in the supplementary guidance. The council rejected the call by Swindon FoE for a wider exclusion zone to be established on the grounds of the alleged health effects of EMF, stating that scientific evidence did not justify such an approach. However, they did take the unease relating to EMF as a justification for promoting the policy from the supplementary guidance to the local plan. They also stated health grounds to be one basis for the exclusion zone as proposed, along with "access, amenity and public acceptability". The council described its approach as "precautionary", as suggested by Swindon FoE. The council made these arguments at the inquiry held into the plan during its defence of its proposed changes to the deposit draft (TBC 1995a).

The inquiry inspector accepted the council's response to Swindon FoE, though she recommended two main alterations: that health reasons should not be used as justification for the exclusion zone and that curtilages should not be included in the policy. Her main concern was with the "general amenity of dwellings", which justified a reasonable distance being set between them and OHLs (TBC 1995b). These recommendations were accepted fully by the council.

Following the inquiry, however, representations were made by NG against the newly proposed policy. They objected to the remaining mention of the possible health effects of EMF, arguing that the lack of supporting scientific evidence made any such references redundant and called the exclusion zones arbitrary and unnecessary. An objection was also brought by the House Builders' Federation. However, the council (now SBC) maintained its position, and the policy passed to adoption in the form recommended by the inspector (SBC 1999). Figure 6.1 shows key elements in the development of the policy.

6.4 KEY POLICY ISSUES

The council's OHL policy is concerned solely with restricting the construction of new dwellings within a certain distance of existing OHLs. It only refers to development near 132 kV OHLs and not those of higher voltage, suggesting that the policy has site-specific planning considerations in mind. One particular OHL in the borough is, in fact, the focus of the policy.

A 132 kV OHL was constructed around the northwest edge of the town of Swindon in 1972, routed in such a way as to avoid the built-up area. However, the growth of the town has since extended up to and beyond this OHL, so that one stretch of the line is now well within the urban limits. The first housing to be built in its vicinity, in the early 1980s, made little allowance for the presence of the OHL, with some dwellings being built directly underneath the conductors (Figure 6.2). Proposals for the residential development of 600 ha of farmland extending across and beyond the route of the OHL were first put forward in 1986, followed

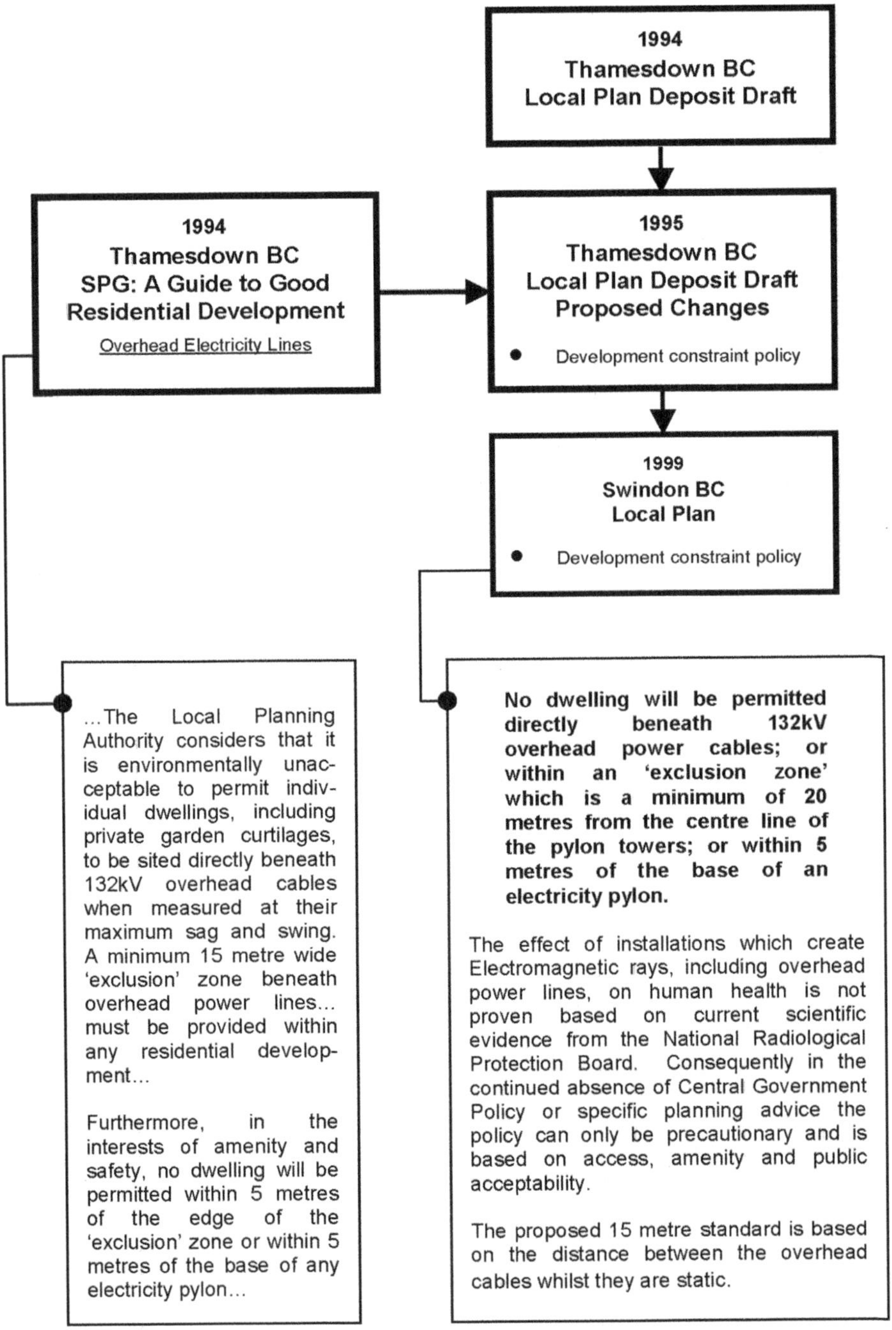

Figure 6.1 Policy development by Swindon Borough Council

Figure 6.2 1980s' housing development beneath a 132 kV OHL, Swindon

Figure 6.3 1990s' housing development and exclusion zone alongside a 132 kV OHL, Swindon

by a masterplan in 1989. The current plan confirms this "Northern Development Area" as the principal focus of growth for the town during the plan period.

This proposed major development was initially resisted by the council and was the subject of a large number of objections focused on the scale and rate of growth and on the likely pressure upon existing services and infrastructure (TBC 1987):

> ... there was this sense of feeling we've all had enough, but we're still getting it; the Northern Development is just another phase in the continued growth of Swindon, and we're fighting all the time new greenfield sites being committed for development ... it really has been quite excessive (Swindon FoE A 2001).

There have been, therefore, deep-seated local concerns about the unusually rapid growth of Swindon. This experience of urban development appears to have encouraged a relatively high degree of environmental interest and activism, such as the formation of the Environmental Forum referred to below (SBC Planning Officer B 2001).

A public inquiry was held into the proposals for the Northern Development Area in 1987, the outcome of which led to a sense of inevitability that the scheme would proceed. Attention therefore turned away from outright opposition to questions of the design of the development and the possibility of compensatory measures (SBC Planning Officer A 2001).

Within this context, Swindon FoE raised the particular concern of the planned location of new housing and a primary school in proximity to the OHL. Attention was focused on the perceived health risk of EMF. One member took on an energetic campaigning role in direct response to the plans, looking into the results of relevant research on EMF and publicising the concerns that she felt (Swindon FoE B 2001):

> ... she ... was particularly horrified about the school, and already ... houses had been built underneath existing overhead power lines ... so there was then a major lobby to try and get the school moved in the masterplan; unfortunately we were totally unsuccessful (Swindon FoE A 2001).

Results of research carried out in Sweden, implying that OHLs were associated with increased cancer risks, were widely reported at the time and stirred public anxiety about the development:

> It was starting to raise concern particularly with people who were already living under power lines, and what was this going to mean for the value of their house, and obviously concern about what effect it might have on their children's health, and ... I think it was a little bit of a flash in the pan as far as the media were concerned, it got a lot of concentrated publicity both locally and nationally ... it definitely did raise awareness (Swindon FoE A 2001).

A local Environmental Forum provided an opportunity for these issues to be debated. This Forum had been set up by the council in 1985 with representation from local environmental interest groups, including Swindon FoE, and some local businesses. Officers also participated in its meetings, and it was treated by the council as a means of public consultation (SBC Planning Officer B 2001). Local planning matters were addressed by a planning working group of the forum, one of the aims of which was "to encourage best practice by developers in order to reduce harmful environmental impacts of development" (Thamesdown Environmental Forum 1996); their work specifically had the Northern Development Area in mind. Guidelines produced by the group in 1992 included recommendations about housing development near OHLs, stating that increasing evidence linking EMF to cancer should lead builders to adopt a precautionary approach and maximise the distance between housing and OHLs, possibly by as much as 50 m (Thamesdown Environmental Forum 1996).

Simultaneously, the council was preparing its own supplementary guidance, "A Guide to Good Residential Development", also in direct anticipation of the planned Northern Development Area (SBC Planning Officer A 2001). The inclusion of the section on development close to OHLs derives partly from a perception that dwellings previously built close to the line were poorly designed and also reflects the public anxieties at the time, as expressed by Swindon FoE and the Environmental Forum. It appears that some officers were keen to incorporate the Forum's suggestions into the guidance, having "environmental credentials" of their own (Swindon FoE A 2001).

When Swindon FoE later objected to the non-inclusion of this section of the supplementary guidance in the local plan, they were, therefore, continuing to press for longer-standing aims and could appeal to the existence of guidance which already went some way to expressing those aims.

Swindon FoE's objection to the deposit draft on this matter was part of a wider campaign to influence the content of the plan. They followed the entire plan preparation process, aiming at changing the text of over 60 policies in all (Swindon FoE A 2001). This is indicative of their strategy of seeking to achieve environmental objectives through engagement with the local planning system. Swindon FoE has built up a relative strength and body of expertise and has deliberately turned its attention to shaping local policy:

> … because there were some … what you could call old campaigners, they weren't satisfied with … some of the hands on activities that other groups were doing, they started to get involved in … politics of changing policy, which is really quite hard to do, it takes quite dedicated people (Swindon FoE A 2001).

As the council had already adopted the OHL section of the supplementary guidance as a form of policy, planning officers proved amenable to its promotion to the local plan. However, when preparing the guidance, they had specifically looked into the question of alleged health risks arising from EMF exposure; consultations with the National Radiological Protection Board (NRPB) at that stage had led them to feel that it was impossible to make any reference to EMF in their recommendations for development near OHLs (TBC 1993). Officers did not therefore fully support the policy as suggested by Swindon FoE; they were also aware that other local authorities had failed when trying to take a restrictive approach:

> ... what we included in that policy was as far as we could go ... there wasn't any support anywhere else, and there were other authorities that were seeking a wider corridor beneath the lines, but they hadn't been supported ... It was one of the Yorkshire authorities, I think, who were seeking a much wider width, 50 metres ... (SBC Planning Officer A 2001).

Swindon FoE's call for a wider exclusion zone was therefore rejected and stress was instead laid in the policy on amenity rather than possible health effects. However, the council's continuing unease with the proximity of new dwellings to the OHL was demonstrated by the fact that, at one point, investigations were carried out into the possibility of relocating the line away from housing altogether (SBC Planning Officer C 2001). The adopted policy has been implemented in relation to recent housing developments, with the creation of open space along certain lengths of the OHL (Figure 6.3).

The handling of the perceived health effects of EMF continues to be an area of difficulty for the council as it undertakes review of the local plan; for instance, following the opening of the school, renewed anxieties were expressed about its proximity to the OHL, especially in the light of fresh media reports implying EMF-related health risks. Officers themselves demonstrate uncertainty on how to respond to these concerns: "We are struggling for advice on this issue" (SBC Planning Officer D 2001); "I think this local authority, and presumably many others, would welcome some sort of statutory guidance" (SBC Planning Officer C 2001).

6.5 CONCLUSION

The urban growth of the town of Swindon has, as elsewhere, led to the local planning issue of the encroachment of residential areas on OHLs that were originally sited to avoid built-up areas. However, the scale of Swindon's expansion has been such that this issue has affected a major development zone. The rapid and continuing rate of growth in an otherwise rural area has also heightened sensitivity to the environmental effects of development. This has encouraged environmental interest and activism and has made the council receptive to local concerns.

Within this context of urban expansion and environmental awareness, a number of more specific features have been instrumental in the formation of the OHL policy:

- the location of the 132 kV OHL around one section of Swindon's perimeter became a consideration in the design of the Northern Development Area;
- media reports on the alleged health effects of EMF raised public and activist anxieties about the possible effects of OHLs;

- the planned proximity of housing and a primary school to the OHL generated particular local concerns;
- campaigning by an individual member of the Swindon FoE raised the profile of these concerns;
- Swindon FoE's tactic of intervening in plan preparation led them to focus on the OHL policy, among others;
- the Environmental Forum and its Planning Working Group provided a channel for policy ideas to be advanced to the council;
- officers' concern for local environmental quality made them responsive to Swindon FoE's representations, though their awareness of NRPB advice and the experience of other authorities made them unwilling to follow these representations fully.

The components of policy formation by SBC are depicted in Table 6.1. The interrelation of the principal components and the progression of policy through the development plan process (DPP) are summarised in Figure 6.4.

ROTHERHAM METROPOLITAN BOROUGH COUNCIL

6.6 INTRODUCTION

Rotherham Metropolitan Borough Council (RMBC) is a unitary authority in South Yorkshire, immediately to the east of Sheffield. It is dominated by the town of Rotherham, which is set in hilly countryside. The area has an industrial past that is based on coal mining and steel making, but serious decline of these activities has led to high levels of unemployment and deprivation and large areas of dereliction. The authority has recently adopted its first unitary development plan (UDP) (RMBC 1999); this case study was undertaken during the period following adoption.

The plan's primary emphasis is on the economic regeneration of the borough through the reuse of former industrial land, in line with regional planning guidance (RPG) (DoE 1989, Government Office for Yorkshire and Humberside 1996). Five major regeneration areas are identified; a considerable number of sites are also allocated for new housing. This strategy is seen as a means of environmental improvement, which will, in turn, stimulate further investment as well as improve the overall quality of life of residents.

There is also some emphasis on the alleviation of social disadvantage. It is envisaged that the strategy of regeneration and improvement of the urban environment will contribute towards this goal, along with other more specific measures targeted at particular aspects of poverty. One such aspect is ill health, borough levels of which are above the national average.

Table 6.1 Components of policy formation by Swindon Borough Council

BROAD CATEGORIES		WITHIN-CASE COMPONENTS	CROSS-CASE COMPONENTS*
INFLUENCING FEATURES	GEOGRAPHICAL, ECONOMIC AND SOCIAL FEATURES	Reports on Swedish EMF research	Media reports on EMF research
		Structure plan allocations	Regional development pressure
		Northern Development Area	Current major housing development
		Sensitivity to effects of recent and planned development	Heightened public environmental awareness
	HVDT INSTALLA-TIONS (AND TELECOMMUNICA-TIONS)	132 kV OHL through Northern Development Area	Installations near existing development
			Installations near proposed development
	POLICY FRAME-WORK	Supplementary Planning Guidance	Supportive local policy instruments
		Precautionary approach	Precautionary principle
		Consultation with NRPB	NRPB advice
	PLANNING EXPERIENCE	Awareness of other LPAs' difficulties in advancing policy	Knowledge of other LPAs' experience
	PLAN THEMES	Environmental aspects of development	Protection of environmental features
ACTORS	PLANNING OFFICERS	Concern for local environmental quality	Strongly supportive of policy
	INTEREST GROUPS	Swindon FoE: group and individual activism	Supportive of/promoting policy
			Study of EMF research
	PUBLIC FORUM	Environmental Forum; Best Practice Guide	Public forum advancing policy ideas
	PUBLIC	Anxieties regarding develop-ment near OHL	EMF health/amenity concerns
	ELECTRICITY INDUSTRY	Objections to modified policy	Opposing aspects of policy
	HOUSE-BUILDING INDUSTRY	Objections to modified policy	Opposing aspects of policy
DPP	REPRESENTATIONS	Swindon FoE suggesting policy (local plan strategy)	Supporting HVDT-related policy
		NG and House Builders Federation objecting to policy	Opposing HVDT-related policy
	LPI/EiP	Inspector recommended alterations	Inspector/Panel recommended alterations
POLICY CONTENT	OVERALL THEMES	Dwellings near OHLs	Residential area protection
	EFFECTS OF HVDT INSTALLATIONS ON:	Public acceptability	Levels of public concern
		Amenity	Amenity
		Human health	Public health (EMF)
		Safety	Public safety
		Dwellings	Residential, etc. areas
	MEASURES TO OFFSET EFFECTS	Exclusion zone	Distancing new development from existing OHLs

* These refer to components defined for the case studies as a whole; see Section 9.2.1.

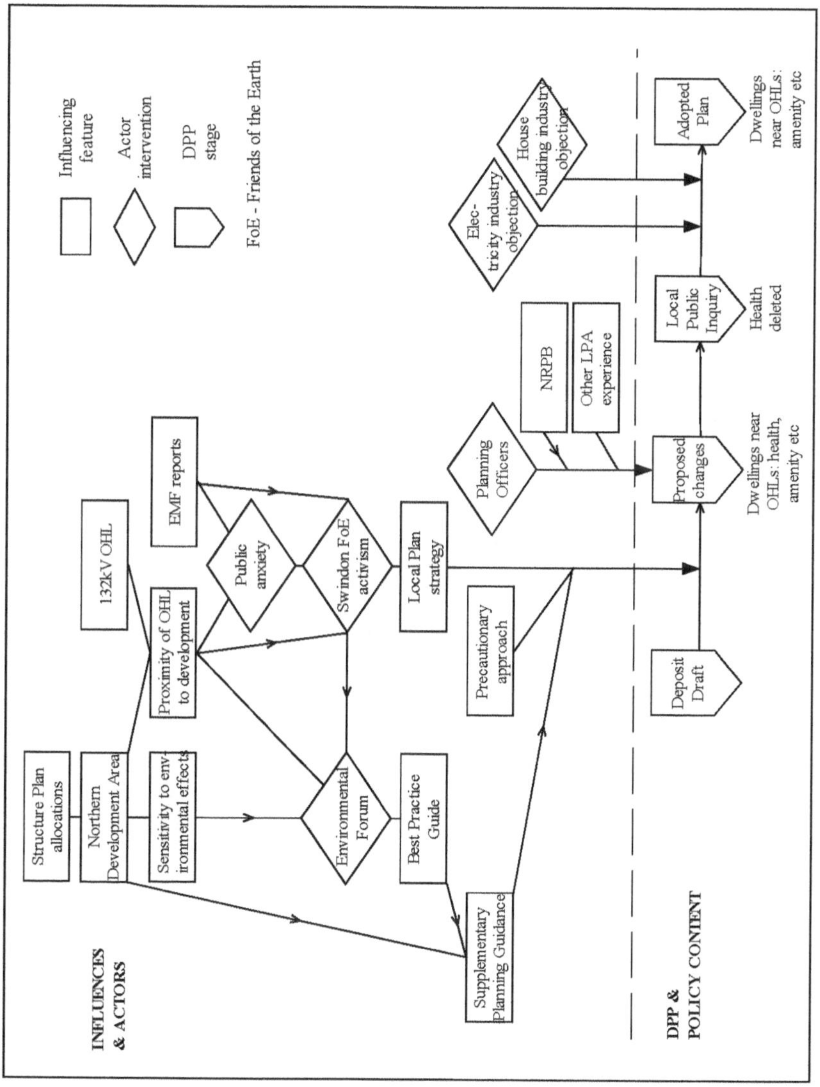

Figure 6.4 The process of policy formation by Swindon Borough Council

6.7 ARTICULATION OF POLICY

Policies referring to HVDT installations appear at every stage in the preparation of RMBC's plan, from the issues paper (1991) onwards. Two main policies recur, both in the section on "Utilities"; they are summarised in Figure 6.5 (RMBC 1995, 1999). A utilities' installations policy deals with the effects of utilities in general, though it is geared principally towards HVDT installations. It is mainly concerned with the visual impact of installations; specific reference is made to OHLs, substations, and so on. The policy recognises the limited role that the LPA has in the placing of these, and calls on the utilities to adhere to statutory responsibilities in minimising visual effects. Residential and environmentally sensitive areas are the main focus of concern. A reference is also made to possible safety issues arising from vandalism. This policy went unchallenged throughout the plan's preparation and remained unchanged from the deposit draft stage onwards.

A development constraint policy deals with development close to existing OHLs. It initially sought to establish a "stand-off zone" along the length of OHLs within which development "with a significant level of human occupation" would not be permitted. The earliest versions of this policy stated this to be "in the interests of health, safety and local amenity" (RMBC 1991, 1994) understood as follows:

- health: possible (cancerous) effects of EMF;
- safety: hazards arising from the swing or drop of conductors;
- local amenity: undefined.

The second policy was the subject of many representations and was changed considerably throughout plan preparation. By the deposit draft stage, the reference to health had been dropped, and the stand-off zone had been reduced from 100 to 50 m wide (RMBC 1995). However, the policy was still justified with reference to public concern about possible health risk. There was recognition of the NRPB's advice that no risk has been clearly established, but the policy was still upheld as a precautionary approach. Support for the policy was sought from a similar Canadian approach and from the electricity industry's own guidelines on safety clearance zones.

The early changes that were made to this policy were a response to objections made by the electricity and house-building industries, who asserted that the council's health concerns could not be substantiated and did not justify a buffer zone along the length of OHLs. These objections were pressed further at the inquiry held into the plan in 1996. For example, NG claimed that the council's precautionary approach was unjustified and that any commitment to the precautionary principle should be based on "fact, not fantasy …; precipitate action on the basis of inadequate evidence is the wrong response" (Her Majesty's Government 1994, p.32).

The council defended its position partly with reference to RPG, which states that new OHLs "should, as far as possible, avoid residential areas and should not be sited in such a way that overhead lines would be required to pass over existing dwellings" (Government Office for Yorkshire and Humberside 1996). The council also referred to a local planning appeal concerning an application to construct housing close to an existing OHL, at Bramley, Rotherham. It had refused the application; the appeal inspector agreed with the council, voicing his concern about the visual effect of the proposals, stating "there would be a disturbing sense of constraint." During the inquiry, however, the council accepted the deletion of all references to health risk in the policy, leaving just safety and amenity reasons as justification. Moreover, the inquiry inspector recommended complete rewording of the policy, replacing the notion of a fixed-distance stand-off zone with a more discretionary consideration of the possible effects of OHLs on proposed nearby development.

The inspector's recommendations were mostly accepted (Figure 6.5). The final version of the policy represents a marked change in emphasis from health concerns to the amenity effects of OHLs, particularly visual impact and noise; a reference to safety is however retained.

Some support for the policy had been received from a residents' association expressing amenity interest, and from English Nature supporting the policy's protection of wildlife. The Countryside Commission had also referred the council to its position statement on OHLs (Countryside Commission 1994); this led to the possibility of undergrounding or re-routeing existing OHLs being introduced into the policy at the stage of proposed changes to the deposit draft (RMBC 1996).

6.8 KEY POLICY ISSUES

6.8.1 Protection of public health and safety

One of the initial aims of the development constraint policy was to avert possible health risk resulting from EMF exposure. The policy was drafted in response to the concerns of elected members, who were in turn responding to public anxieties. Questions were raised, for example, over the proximity of a children's play area to an OHL (RMBC Elected Member 2001).

It appears that public and member attitudes were being shaped by media reports on the possible harmful effects of EMF. These reports were current throughout the 1990s when the plan was being prepared; for example, the council made reference to a 1996 television documentary *Electricity and Cancer* during its defence of the policy at the inquiry stage. Planning officers reflected these concerns:

> ... we were also getting pressure from our local members, because they were picking up the press reports ... There was stuff coming out from Bristol University, certain studies that had been done there seemed to prove that there might be a link ... So we devised a policy ... trying to create a development free barrier along the major routes of cables ... (RMBC Planning Officer A 2000).

1995
Rotherham MBC
UDP Deposit Version

- Utilities' installations policy
- Development constraint policy

1999
Rotherham MBC
UDP Adopted Version

- Utilities' installations policy
- Development constraint policy

Environmental Impact of Service Installations

The Council will seek to ensure that the utility companies and agencies avoid or where this is not possible, minimise the adverse landscape and environmental impacts of transmission lines, installations and other similar apparatus.

...certain service installations... such as electricity switching stations... and overhead power lines, can be visually intrusive, particularly in residential areas and areas of environmental sensitivity. They can also be a potential source of danger if they are not kept secure from vandalism and trespass... reliance is placed upon the utility companies and agencies to adhere to their statutory environmental obligations.

High Voltage Power Lines

The Council will not permit new development directly beneath high voltage (132kV and above) power lines in the interests of safety and local amenity. Furthe-rmore, the Council will discourage development with a significant level of human occupation within 25 metres of any high voltage... power line.

There is growing public concern about the possible health effects of electric and magnetic fields associated with major overhead power lines...
The National Radiological Protection Board advise that evidence of harmful effects on people... is not clearly established...
In Canada... power lines may not be sited within 50 metres of dwellings.
...a precautionary approach is prudent. Therefore, the Council proposes a 25 metre 'stand-off zone'...

Development close to High Voltage Power Lines

Applications for development close to overhead power lines should take into account the effect of the transmission towers and cables in the vicinity of the site on the amenity of the occupiers of the proposed development.

For reasons of safety and amenity the Council will generally discourage development close to overhead power lines...
...dwellings located close to power lines could be dominated by them and views from the properties significantly impaired. Further, under certain conditions some power lines emit noises...
...the Council will seek to discourage development for housing, schools, hotels, industry and all other forms of land use that would result in people spending a significant proportion of their time in a visually poor environment. The exact distance considered appropriate will be dependant upon a combination of factors...

Figure 6.5 Policy development by Rotherham Metropolitan Borough Council

This issue should be seen in the context of a wider council concern for the health of the population of the borough as a whole. Rotherham has high incidences of ill health, and the council has set itself a "Health for All Strategy" that the plan explicitly incorporates. The possibility of averting alleged health risk by preventing development near OHLs provided an opportunity for planning to contribute towards this strategy:

> Health issues are something that Rotherham has got to the fore … the thought was maybe we could do something on this one, and be seen to be doing something (RMBC Planning Officer A 2000).

Health concerns relating to EMF have been expressed by several other authorities in South and West Yorkshire in their policies. Communication took place between council officers and members and their counterparts in some of these other authorities with respect to their policies. In particular, Calderdale Metropolitan Borough Council preceded RMBC in attempting a similar approach of establishing a buffer zone alongside OHLs, and thus provided an example for RMBC to follow. However, Calderdale did not succeed in adopting the policy in the form that it had proposed, in the face of objections from the electricity industry. The inspector of the inquiry held into the Calderdale plan concluded that the council's health concerns were unsupported by scientific evidence and that there was therefore no justification for a protection zone; furthermore, a zone of this kind could cause unnecessary concern among those already living close to OHLs and blight their property (Calderdale Metropolitan Borough Council 1994). RMBC officers kept in touch with the progress of Calderdale's policy, and, as a result of Calderdale's difficulties, sought to modify their own approach. However, they were bound by the persisting commitment of their own elected members to the development constraint policy:

> … it became quite obvious as it continued and we watched what was happening … that we probably weren't going to get anywhere, but we were guided by our political masters … they insisted that the policy stayed in (RMBC Planning Officer A 2000).

The assumption among officers was that the electricity industry's strong opposition to any policy reference to EMF was for legal and financial reasons:

> Well I can understand why they were adamant against it … I'd assume that the main fear was where lines already pass over or close to existing properties, the possibility of claims coming in, damages, etc., or compensation … (RMBC Planning Officer A 2000).

In seeking to defend the policy, officers called on pieces of evidence that were in themselves further motivating forces for the policy. The appeal to the precautionary principle was seen in this way, as there was a genuine feeling that it should be applied in the case of EMF-related health concerns.

> Our position was always one where we felt we should adopt the precautionary principle which the government is keen to promote and encourage; it seemed a bit strange to us frankly that they didn't see in this circumstance a precautionary principle to apply; we felt that the

> jury is still out to an extent, and that until there was no doubt at all we should operate a precautionary principle ... (RMBC Planning Officer B 2001).

Similarly, research undertaken by officers provided them with international examples of buffer zones and directed them towards a more reasoned justification of the 50 m zone being advocated. Being put on the defence therefore served to increase their commitment to the need to protect public health. There was, however, some feeling that insufficient government advice was available: "National guidance would have saved a lot of trouble" (RMBC Planning Officer A 2001).

The references to safety issues in the policies, relating for example to the drop of conductors, are also in line with a concern for health. It is possible that they were included in the development constraint policy partly with the aim of substantiating the stand-off zone.

6.8.2 Minimisation of visual and amenity effects

Concern for the visual effects of HVDT installations is expressed both in the utilities' installations policy and also in the adopted form of the development constraint policy. In the case of the development constraint policy, visual concerns were introduced to substitute for health concerns that were removed at the inquiry. Along with this, the development constraint policy also raises concern about the effects of installations on amenity; the adopted version mentions noise effects as an example.

These issues correlate closely to those raised in the Bramley planning appeal to which the council referred at the inquiry; in that appeal, both the visual and noise effects of an existing OHL were cited by the inspector. The inquiry inspector therefore followed the appeal inspector in using the same grounds for supporting a modified form of the development constraint policy. Visual concerns should also be seen in the context of broader aims of the plan to preserve the visual quality of the borough and to minimise the impact of developments on the landscape:

> The quality of the landscape, both urban and rural, is important not only to the quality of life of its residents but also for attracting economic development and visitors ... Priority needs to be given to the protection and enhancement of the landscape ... (RMBC 1999).

6.8.3 Residential proximity to HVDT installations

Both the above areas of concern – health and safety and visual and amenity effects – extend from concerns about the relationship between HVDT installations and other types of development. The spatial distribution of these different land uses in the borough is an important consideration in understanding the formation of RMBC's policies.

The borough contains a high concentration of existing HVDT installations; these include four NG-owned substations and associated OHLs and equipment built to serve the steel-making industry (Figure 6.6). Much of this infrastructure lies within urban areas, often close to residential estates, and is highly visible. For example, the 400 kV Brinsworth substation is situated on the edge of a housing estate and has six OHLs radiating from it across hilly terrain, with many towers visible on the skyline (Figure 6.7). There is, therefore, a strong public awareness of the presence of HVDT installations in some parts of the borough.

Figure 6.6 HVDT installations near steelworks, Rawmarsh, Rotherham

Figure 6.7 HVDT installations near housing, Brinsworth, Rotherham

This sets the scene for seeking the minimisation of the "adverse landscape and environmental impacts of transmission lines ..." as stated in the utilities' installations policy. It also provides the background to concerns regarding residential proximity to equipment. There has been a history in the borough of installations being located close to housing:

> We've got a situation where the population lived very close to the sources of industry ...; if you've got steelworks with massive power lines coming in, they're coming in reasonably close to housing as well. There were worries, because the industry's gone but the housing in many cases is still there (RMBC Planning Officer A 2000).

It is in this context of a dominant presence of equipment close to residential areas that policies designed to mitigate the various effects of HVDT installations should be understood. Press reports raising questions about the health effects of OHLs are likely to find fertile ground in this setting. Although the development constraint policy deals with the consequences of potential new development near OHLs, the existing pattern of development is a contributory factor to this policy. In other words, concern for present residents near various installations finds expression in the opposition to future development adjacent to OHLs.

Future residential development in the borough is dealt with in the plan as part of the council's regeneration strategy. The plan identifies a large number of mainly brownfield sites for new development, including housing. The plan policy, which allocates new sites for residential use, makes a cross reference to the development constraint policy, indicating that the relationship of new housing to HVDT installations is an expected planning issue within the scheme of regeneration. Indeed, some of the allocated sites are within the vicinity of existing OHLs.

A link can also be found between the policies in question and the plan's central theme of regeneration in a wider sense. Those policies dealing with the adverse effects of certain developments, such as HVDT installations, form part of the overall aim of environmental improvement, which is designed both to improve the overall quality of life in the borough and to increase the attractiveness of the area to inward investment.

6.9 CONCLUSION

RMBC's first UDP sets itself a strategy of economic regeneration for the borough and links this to associated needs of alleviating social deprivation and bringing about environmental improvement. This strategy provides a context within which the plan's policies on HVDT installations can be placed. Certain aspects of the steps towards regeneration contribute towards the formation of these policies:

- the aim of improving the physical environment, both to improve the local quality of life and to encourage investment, raises the issue of the visual and amenity effects of OHLs;
- the allocation of brownfield sites to new housing raises the question of the inter-relation of future housing and OHLs on some of these sites;
- the priority on public health raises concern with the perceived, possible health effects of EMF.

For these issues to be raised within this context of regeneration, however, a number of other features have been instrumental:

- the borough has a high density of HVDT equipment, especially within its urban areas;
- similar policy concerns have been raised by other nearby local authorities, particularly in relation to EMF and buffer zones;
- national media reports have increased anxieties about the alleged effects of EMF among the public and elected members;
- the council has had recourse to wider policies and practice which it has perceived to be supportive of its position: regional policy, the precautionary principle and international examples;
- the development control case of Bramley has raised the question of the proximity of development to existing OHLs in a specific, local way.

In conclusion, RMBC's polices on HVDT installations can be interpreted as the attempt of a local authority to deal with the adverse effects of a concentration of highly visible equipment, particularly in the urban setting, at a time when environmental and social improvements are seen as integral to the future of the borough. External influences, notably media reports and the concerns of other authorities, have led policy partly in the direction of alleged health effects; the council's emphasis on improving health has made policy formation particularly responsive to these worries. Local planning considerations, especially anticipated difficulties on new housing sites, have contributed to the attempt to ensure stand-off zones along OHLs. Objections from the electricity industry to the council's approach turned policy more towards visual and amenity issues, but these also have their origin in a wider concern for environmental quality and improvement. The components of policy formation by RMBC are depicted in Table 6.2. The inter-relation of the principal components and the progression of policy through the DPP are summarised in Figure 6.8.

Table 6.2 Components of policy formation by Rotherham Metropolitan Borough Council

	BROAD CATEGORIES	WITHIN-CASE COMPONENTS	CROSS-CASE COMPONENTS*
INFLUENCING FEATURES	GEOGRAPHICAL, ECONOMIC AND SOCIAL FEATURES	Media EMF reports	Media reports on EMF research
		Industrial decline and dereliction	Economic decline/poor environmental conditions
	HVDT INSTALLATIONS (AND TELECOMMUNICATIONS)	High density of installations in urban area	High density of installations
		HVDT installations near existing and proposed housing sites	Installations near existing development
			Installations near proposed development
	POLICY FRAMEWORK	Supportive regional guidance	Supportive regional/structure planning policy
		Other Yorkshire authorities' policy	Comparable neighbouring authority policy
		Appeal to precautionary principle	Precautionary principle
		Canadian policy examples	International precedents
		Recognition of NRPB advice	NRPB advice
		Electricity industry's safety clearances	Electricity industry guidelines
	PLANNING EXPERIENCE	Bramley planning appeal	Electricity installation development control cases
		Knowledge of nearby authorities' experience	Knowledge of other LPAs' experience
	PLAN THEMES	Regeneration strategy; environmental protection	Regeneration/environmental improvement
		Health strategy	Public health
ACTORS	PLANNING OFFICERS	Seeking policy examples	Study of EMF research
		Awareness of difficulties faced by Calderdale MBC	Caution against inclusion of EMF health concerns
			Communication with officers from other authorities
	ELECTED MEMBERS	Members responding to public anxieties	Strongly supportive of policy
			EMF health/amenity concerns
		Communication with members from other authorities	Communication with members from other authorities
	PUBLIC	Public EMF concern, e.g. play area	EMF health/amenity concerns
	ENVIRONMENTAL BODIES	Supporting proposed policy	Supporting aspects of policy

	BROAD CATEGORIES	WITHIN-CASE COMPONENTS	CROSS-CASE COMPONENTS*
DPP	ELECTRICITY INDUSTRY	Objections to proposed policy	Opposing aspects of policy
	HOUSE-BUILDING INDUSTRY		Opposing aspects of policy
	REPRESENTATIONS	Environmental, etc. bodies supporting policy	Supporting HVDT-related policy
		NG and house-building bodies objecting to policy	Opposing HVDT-related policy
	LPA DEFENCE	LPA defending policy at LPI	Strongly defending policy against objections
	LPI/EiP	Contestation	Contestation
		Recommended rewording of policy	Inspector/Panel recommended alterations
POLICY CONTENT	BROAD THEMES	OHLs and nearby development	Residential area protection
	EFFECTS OF HVDT INSTALLATIONS ON:	Visual quality: residential areas	Visual quality: urban areas
		Public anxieties: EMF	Levels of public concern
		Amenity (undefined)	Amenity
		Health: EMF	Public health (EMF)
		Safety: vandalism, hazards	Public safety
		Residential areas: visual impact	Residential, etc. areas
		Noise levels	Noise levels
	MEASURES TO OFF-SET EFFECTS	Reliance on statutory responsibilities of utilities	Reliance on statutory responsibilities of utilities
		Re-routeing existing OHLs	Removal/re-routeing of existing OHLs
		Undergrounding of lines	Undergrounding of lines
		Stand-off zones	Distancing new development from existing OHLs

* These refer to components defined for the case studies as a whole; see Section 9.2.1

NEWHAM COUNCIL

6.10 INTRODUCTION

Newham Council (NC) is an inner east London borough authority, which is densely populated, and has high levels of social deprivation. The borough forms a block on the north side of the Thames, including the dockland area of the Royal Docks and the Lower Lea Valley, which feeds into the Thames. Both of these areas have, in the past, attracted major industries and utilities, some of which were heavily polluting; however, industrial decline and closure of the docks, mainly in the 1970s, have led to large-scale dereliction. Concerted regeneration efforts began in the 1980s, when the Royal Docks came within the remit of the London Docklands Development Corporation (LDDC), an urban regeneration body with some statutory planning functions. This led to the building of the City Airport, infrastructure improvements and a number of continuing commercial and housing projects. Attention has now turned to the renewal of the Lower Lea Valley; an international design competition, launched partly by the council, has produced plans for high-quality residential and commercial development.

The council first adopted a UDP in 1997. The plan's principal aim is the progressive regeneration of the former industrial areas; it finds unparalleled potential in the availability for development of more than 400 ha of land in inner London. Moreover, renewal on this scale is seen as having strategic importance, acting as a "catalyst for the regeneration of East London and therefore the whole of the Thames Gateway" (NC 1997). Overall, "The council's aim is to secure for London a new commercial, educational, cultural and leisure complex of European significance" (NC 1997, p. 1). The plan seeks to assist this process through the designation of "Major Opportunity Zones", for which broad specifications are given and potential land uses indicated. A review of the plan, completed in 2001, reasserts this strategy of regeneration (NC 2001a); Supplementary Planning Guidance has also been prepared to guide development in the Lower Lea Valley.

6.11 ARTICULATION OF POLICY

Policies relating to OHLs appear throughout the preparation of NC's plan, though they show considerable modification during the process. They demonstrate two principal concerns, as summarised in Figure 6.9.

One policy deals with the distancing of new development from existing OHLs. At the 1993 deposit stage, this established "exclusion zones" at least 20 m wide along the length of OHLs within which residential and some other forms of development would be discouraged. This approach was justified as a precautionary measure against noise and the effects of EMF (NC 1993).

Objections to this policy were made by NG, who argued that the reference to the precautionary principle and the establishment of an exclusion zone were supported neither by the NRPB advice nor by the recommendation of the inspector at a recently held plan inquiry (Calderdale Metropolitan Borough Council 1994). The council accepted these representations and modified the policy to state that built development would be separated from OHLs, but not to any set distance, and that this would be principally on the basis of visual amenity.

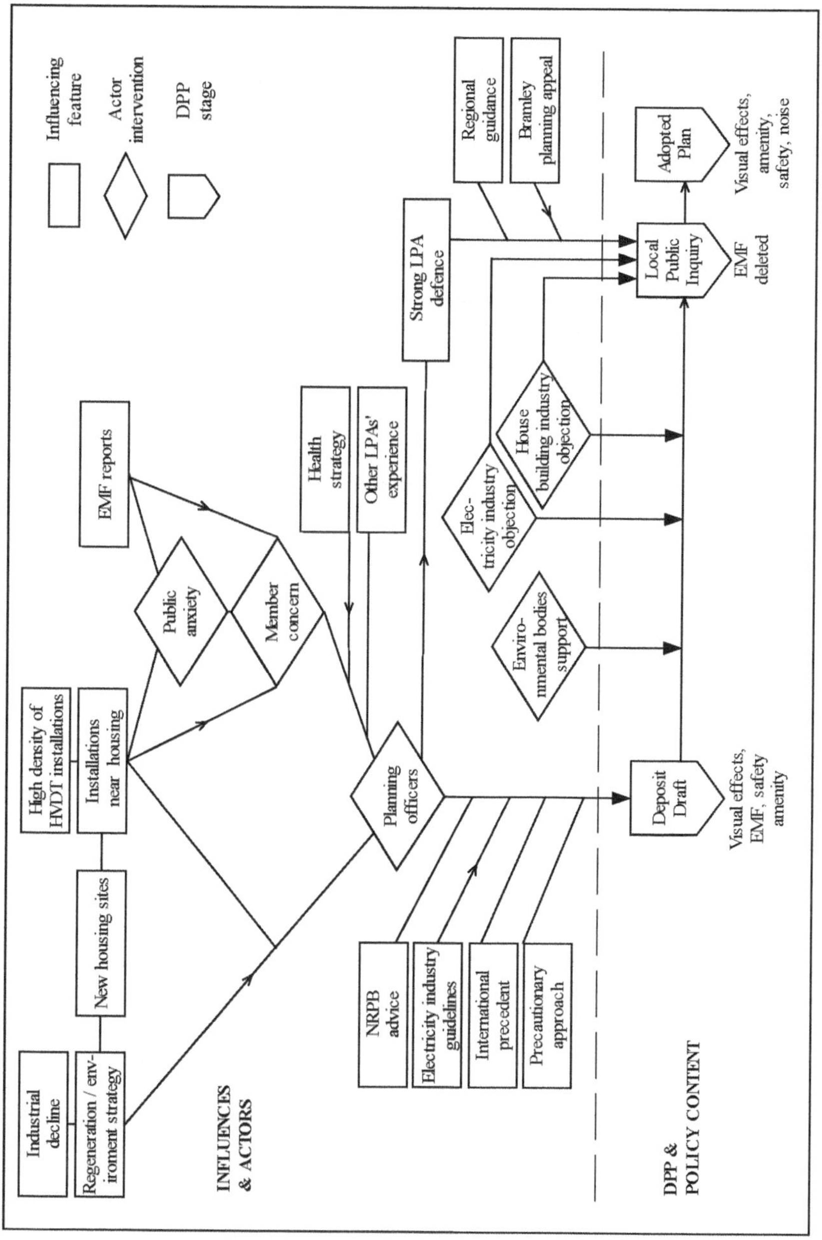

Figure 6.8 The process of policy formation by Rotherham Metropolitan Borough Council

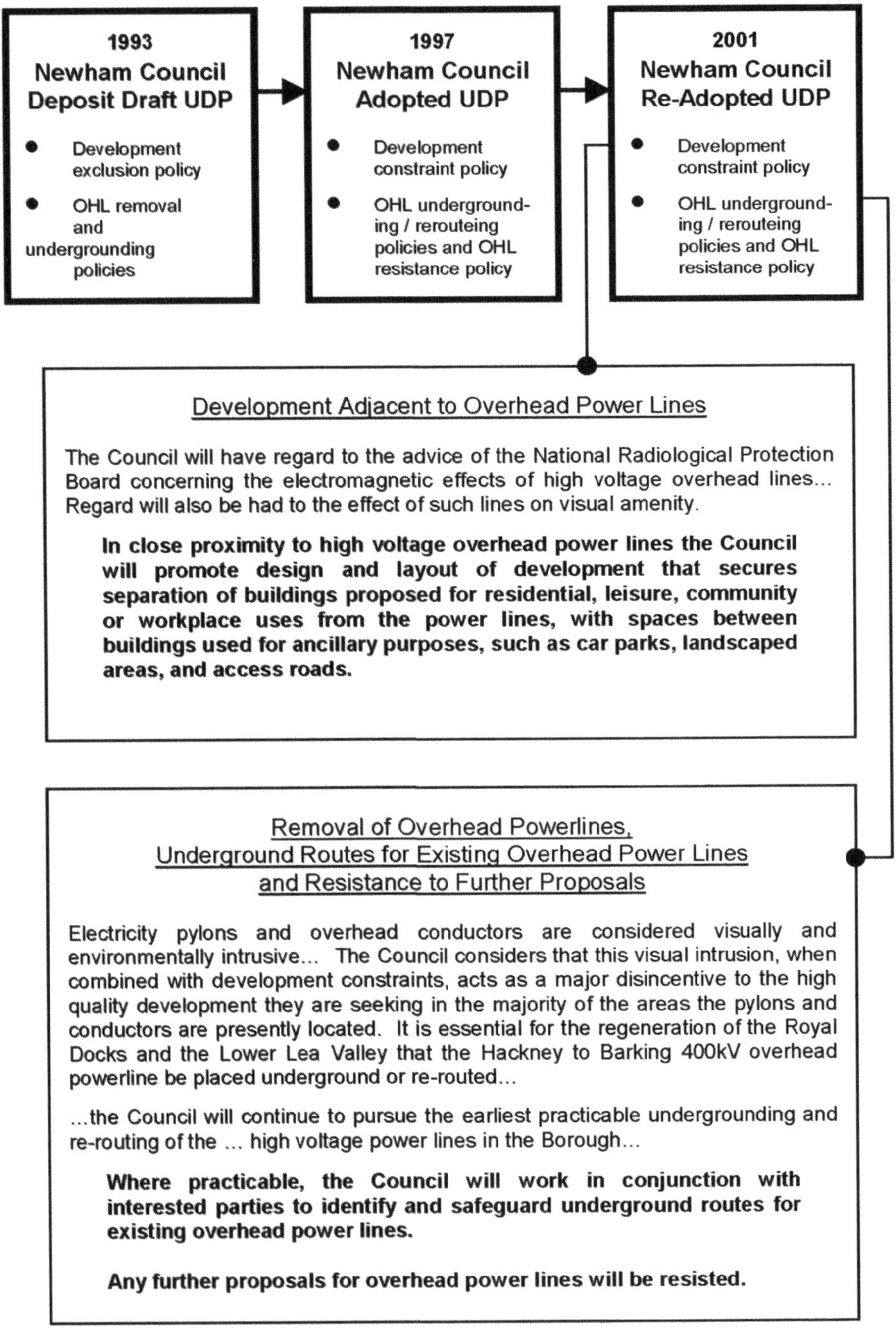

Figure 6.9 Policy development by Newham Council

Two policies seek the removal of existing OHLs in the borough: the 400 kV "Hackney to Barking" line and two 275 kV lines. They are described as a disincentive to development because of their visual intrusiveness and the physical constraints that they impose; their removal is advocated in the interests of regeneration. The plan puts forward undergrounding as a means of achieving the removal of OHLs and proposes to set aside underground routes in order to facilitate this. The re-routeing of OHLs is also suggested as a possible solution. The deposit draft also refers to specific sections of OHL described as "constraining" certain development sites (NC 1993).

NG also made representations in relation to these policies. They objected to

- the aim of removing OHLs for which consent has been granted; in particular, the 400 kV "Hackney to Barking" line had been the subject of a wayleave hearing, as a result of which the Secretary of State had granted a wayleave for the line for a further 15 years;
- the lack of understanding about the practicalities of removing or undergrounding OHLs;
- the application of the same criteria to all OHLs, when the wayleave hearing inspector had judged the circumstances of the 400 kV line to be exceptional.

The council resisted these arguments at the subsequent UDP inquiry, asserting that its aim of making environmental improvements justified its stance. It referred to RPG prepared for the "Thames Gateway", which calls for the need for the area to break out of a cycle of environmental blight and mentions the undergrounding of OHLs as an example of setting new standards.

The inquiry inspector gave broad support to the case made by the council for seeking removal of OHLs, though recommended that much of the policy wording would be more appropriately written as reasoned justification (NC 1995). In consequence, the 1997 adopted plan considerably modified the expression of policy, but retained the council's fundamental aims. A separate policy specifically resisting the siting of any further OHLs in the borough was also formulated (NC 1997).

A "rapid review" of the plan was carried out between 1998 and 2001. The policies relating to OHLs were carried forward to re-adoption with relatively little change (NC 2001a). However, the second of the policy concerns, the desired removal of OHLs, was given renewed attention in the Supplementary Planning Guidance prepared for the Lower Lea Valley. This describes OHLs in the area as an eyesore and a restriction on development and aims to work with relevant electricity companies in order to assess the possibility of undergrounding them (NC 2001b). Site-specific references to OHL equipment are also made in the accompanying guidance notes (NC 2001c).

6.12 KEY POLICY ISSUES

6.12.1 Proximity of residential and other development to OHLs

Concerns about the proximity of residential and certain other types of development to OHLs arose initially owing to the closeness of parts of the 400 kV "Hackney to Barking" line to existing housing. In particular, housing had been built in Beckton in the early 1980s with little regard for the line; one transmission tower, for instance, found itself located immediately adjacent to a number of the new properties (Figure 6.10). Local anxiety about the presence of the OHL arose following national media reports in the early 1990s about the possible health effects of EMF exposure. These concerns were strongly expressed by local community groups and in public meetings and received elected member support. The voicing of community feeling coincided with early stages of UDP preparation, and one member, who chaired the council's Environment Committee, made efforts to ensure that health issues associated with OHLs were addressed in the plan (NC former Planning Officer 2001, NC Planning Officer 2001).

Figure 6.10 1980s' housing close to a 400 kV OHL, Beckton, Newham

In the deposit draft of the plan, the policy that responded most explicitly to health concerns was that which proposed development exclusion zones around existing OHLs; this drew upon the precautionary principle as a means of justification and made direct reference to health worries (NC 1993). However, this policy was not dealing with the issue of housing, and so on, already located near OHLs, but seeking to prevent further development of this kind. This can be interpreted partly as an anticipatory measure, and also as a policy constrained by a land-use pattern that is beyond recall. The other OHL policies in the plan, however, *do* challenge the proximity of OHLs to existing housing, by calling for their removal. But these policies are placed entirely within the context of regeneration, as described below; they make no mention of health concerns. Nevertheless, it would appear that EMF issues were partly in mind during their formation, but that there was insufficient confidence that health concerns could be defended in respect of these policies at the inquiry stage of the plan: "We didn't want to obscure the primary aim which was the regeneration ... that was the tactical advice" (NC Elected Member 2001).

6.12.2 Removing disincentives to regeneration

The borough sees itself as having a poor environmental image that acts as a deterrent to inward investment. This is traced back historically to Newham's position just beyond the boundary of the former County of London. Many industries and utilities were prohibited by nineteenth-century legislation from locating within the County because of their polluting effects, yet were required to serve London; they would therefore be placed around its perimeter. The Lea Valley and the Docklands provided advantageous sites and so became the home to many "dirty" industries (NC 1997, NC Regeneration Officer 2001). (The pattern of OHLs in the area can be seen as a consequence of this trend, as they served some of the industrial facilities, and were able to use the industrial corridors as routes into London.) This history of poor environmental standards and practice was later compounded by the dereliction caused by the closure of the docks and the loss of industry. The plan describes this set of circumstances as a "self-reinforcing cycle of environmental blight" which it is imperative to break out of (NC 1997, p. 23). This follows the "Thames Gateway" regional guidance, which includes among the problems of the wider area:

> ... a legacy of environmental degradation: the place where London generated its power and dumped its rubbish, producing derelict sites and a web of overhead power lines ... a lack of ambition and a readiness to accept uninspired development ... (DoE 1995, p. 7).

By contrast, the plan describes a vision of renewal for the Lower Lea Valley and Royal Docks which sees the vast areas of dereliction as a unique opportunity for major development. Here also, it is in line with regional guidance, which identifies the Royal Docks and the Lower Lea Valley as being in a "pivotal position at the interface ... with central London" and encourages high-quality development (DoE 1995, p. 37). The regeneration strategy proposed by the guidance includes "a new environmental standard" aimed to change the image of the area, and the amelioration of conditions that inhibit regeneration.

The removal of OHLs in the borough is seen by the council as an essential element in achieving the desired reversal in past trends. This is illustrated by the deposit draft's comment on the "Hackney to Barking" line, as it runs across one of the plan's major development zones:

> This line presents a partial development constraint, but more significantly it casts a visual and environmental blight that prevents the site realising its full potential. The Council … has pursued action to have the line removed or put underground … (NC 1993, p. 117).

and by this officer opinion:

> … we had miles and miles of wasteland in the Royal Docks, and the main focus of development in London has been in the west of London or increasingly in … Canary Wharf; … investment just wasn't coming into Newham, and obviously the borough's main concern was to try and get rid of anything which acted as a detraction to investment … (NC Planning Officer 2001).

The notion that the removal of OHLs would assist the economic regeneration of the borough can be traced back to the 1980s when district plans produced by the council made specific reference to the "Hackney to Barking" line across the Royal Docks; they described it as unattractive, noisy and a constraint on development, and called for it to be re-routed or placed underground (London Borough of Newham 1980, 1985a,b).

Furthermore, in 1989, the LDDC led an action to try and bring about the removal of this line (NC former Planning Officer 2001). As owners of the land over which part of the line passed, they, along with the council and other landowners, served notice on NG for the termination of their voluntary wayleave agreements. This led, in 1992, to a major wayleave hearing (referred to earlier), by which NG sought to obtain a compulsory, or "necessary", wayleave from the Secretary of State. During the hearing, LDDC argued that the removal of the line, preferably by undergrounding, was essential for the high-quality development envisaged for the Royal Docks, particularly as the line dominates entry points to the area. However, the inspector did not feel that an overwhelming case had been made for the line's removal and recommended the granting of a necessary wayleave, albeit subject to review within a few years. He was critical of the precipitate withdrawal of voluntary wayleaves and the failure to negotiate with NG, particularly by LDDC, who had taken the initiative in the action (Planning Inspectorate 1992). The lead role played by LDDC was partly due to the financial resources at their disposal; the council alone would have been unable to pursue this type of action (NC Elected Member 2001).

The inspector's recommendation that a review of the wayleave should take place appears to have encouraged the council to pursue its aim of undergrounding the OHL by means of the UDP. The policies that were introduced on the subject alluded to the withdrawal of wayleaves by LDDC and the council, suggesting that the case had had a formative effect on policy. The LDDC were also involved in the development of these policies:

> At that time, the LDDC were the Planning Authority for the Royals area ... for development control purposes. We both agreed an approach that these overhead powerlines detracted from the image of the borough and impeded investment ... we identified an undergrounding route in the UDP (NC Planning Officer 2001).

Hence, in versions of the plan pre-dating the LDDC's abolition in 1998, the policies refer to the council and the LDDC jointly pursuing OHL removal. The council had active officer and member support for the policies, as it had done for its role in the wayleave hearing (NC former Planning Officer 2001, NC Elected Member 2001).

To date, no OHLs in the borough have been re-routed or placed underground. However, major projects have continued to be developed in the Royal Docks, including a major exhibition centre in close proximity to the 400 kV line (Figure 6.11). The current understanding of council officers is that environmental and infrastructure improvements are now sufficiently advanced, and pressures for development are so great that the 400 kV OHL in the Royal Docks is not impeding development in the way that was once feared (NC Planning Officer 2001, NC Regeneration Officer 2001). However, OHLs continue to be a concern in the Lower Lea Valley, where the environment remains unattractive, and the proposed development is of a smaller scale than in the Docks; it is felt that OHLs are still a disincentive to the intended creation of residential areas with green spaces and waterways (Figure 6.12). Pressure therefore persists for the removal of OHLs, as shown by the re-adopted plan and Supplementary Planning Guidance (NC 2001ab,c). A masterplan for the area provides for the possibility of a services corridor that could incorporate an underground high-voltage electricity route (NC Regeneration Officer 2001).

6.13 CONCLUSION

Newham is a London borough with major OHLs passing through it; the council has developed two areas of policy in direct response to these installations. Firstly, the concern for the proximity of housing, and so on, to OHLs arose as a result of the closeness of the 400 kV OHL to a specific neighbourhood, thus raising particular issues in a localised manner. Other factors also contributed to the formation of this policy:

Figure 6.11 Development close to a 400 kV OHL, Royal Docks, Newham

Figure 6.12 OHLs in the Lower Lea Valley, Newham

- media reports created anxieties about the possible health effects of EMF;
- local concern was voiced through community forums and elected members;

- representations from the electricity industry led to a shift in policy towards amenity considerations.

Secondly, the objective of removing OHLs, preferably by undergrounding, results from their presence in precisely those areas that the council has set itself the priority of redeveloping. The plan is dominated by its aim of regenerating the borough's former industrial sites; the perception that OHLs standing across these sites are an obstacle to fulfilling this aim has driven the desire to see them removed. The ambitiousness of the vision of regeneration, seen as a means of rectifying East London's past ills, has given particular force to measures aimed at environmental improvement; the OHLs themselves are symptomatic of the industrial activity that has blighted the region. This underlying concern has been brought to policy expression by a number of factors:

- regional guidance provided a supportive policy framework;
- LDDC's initiative in seeking undergrounding of the 400 kV line (reflecting the body's purposefulness and resources) encouraged the council to pursue the same objective;
- the wayleave hearing provided an impetus to the formation of policy and presented the possibility of policy aims being achieved;
- officers and members showed commitment to the desire to see OHLs removed;
- anxieties about the possible health effects of EMF, though not mentioned in the policies relating to regeneration, provided additional motivation for pursuing the removal of OHLs.

The components of policy formation by NC are depicted in Table 6.3. The inter-relation of the principal components and the progression of policy through the DPP are summarised in Figure 6.13.

Table 6.3 Components of policy formation by Newham Council

BROAD CATEGORIES		WITHIN-CASE COMPONENTS	CROSS-CASE COMPONENTS*
INFLUENCING FEATURES	GEOGRAPHICAL, ECONOMIC AND SOCIAL FEATURES	Media reports on EMF research	Media reports on EMF research
		History of environmental blight and industrial decline	Economic decline/poor environmental conditions
		Major land availability in prime location	Potential for major development
	HVDT INSTALLATIONS (AND TELECOMMUNICATIONS)	Major OHLs through residential area and areas of major development potential	Installations near existing development
			Installations near proposed development
	POLICY FRAMEWORK	Thames Gateway RPG	Supportive regional/structure planning policy
		1980s district plans	Supportive local planning instruments
		Wayleave hearing recommendation	Supportive HVDT-related public inquiry decisions
		Precautionary principle	Precautionary principle
		Acceptance of NRPB advice	NRPB advice

BROAD CATEGORIES		WITHIN-CASE COMPONENTS	CROSS-CASE COMPONENTS*
	PLANNING EXPERIENCE	1980s' plans; attempted withdrawal of wayleave	History of HVDT-related policy
		Wayleave hearing	HVDT-related public inquiries
	PLAN THEMES	Major regeneration	Regeneration/environmental improvement
ACTORS	INSTITUTIONS	LDDC: planning role and resources	New institutional arrangements
	PLANNING OFFICERS	Commitment to OHL removal	Strongly supportive of policy
		Lack of confidence regarding defence of EMF concerns	Caution against inclusion of EMF health concerns
	ELECTED MEMBERS	Commitment to OHL removal	Strongly supportive of policy
		Voicing of public EMF concerns	EMF health/amenity concerns
	PUBLIC	EMF health concerns and protests	EMF health/amenity concerns
	DEVELOPMENT CORPORATION	LDDC leading attempted withdrawal of wayleave, and supporting LPA policy	Supporting aspects of policy
	ELECTRICITY INDUSTRY	Objections to proposed policies	Opposing aspects of policy
DPP	REPRESENTATIONS	NG opposing policies	Opposing HVDT-related policy
	LPA DEFENCE	LPA defending regeneration efforts	Strongly defending policy against objections
	LPI/EiP	Contestation	Contestation
		Inspector recommended alterations	Inspector/Panel recommended alterations
POLICY CONTENT	BROAD THEMES	Distancing development from OHLs	Residential area protection
		OHL removal from regeneration sites	Economic regeneration
	EFFECTS OF HVDT INSTALLATIONS ON:	Visual quality	Visual quality: urban areas
		Health: EMF	Public health (EMF)
		Residential, leisure, etc. areas	Residential, etc. areas
			Recreational areas
		Development: physical constraints	Land use
		Regeneration of Royal Docks and Lower Lea Valley	Regeneration efforts
	MEASURES TO OFFSET EFFECTS	Resisting new OHLs	Opposition to new HVDT installations
		Removal of OHLs by undergrounding/re-routeing	Removal/re-routeing of existing OHLs
			Careful routeing of new OHLs
			Undergrounding of lines
		Development exclusion zones	Distancing new development from existing OHLs
		Separation of development and OHLs	Sensitive design of development near OHLs

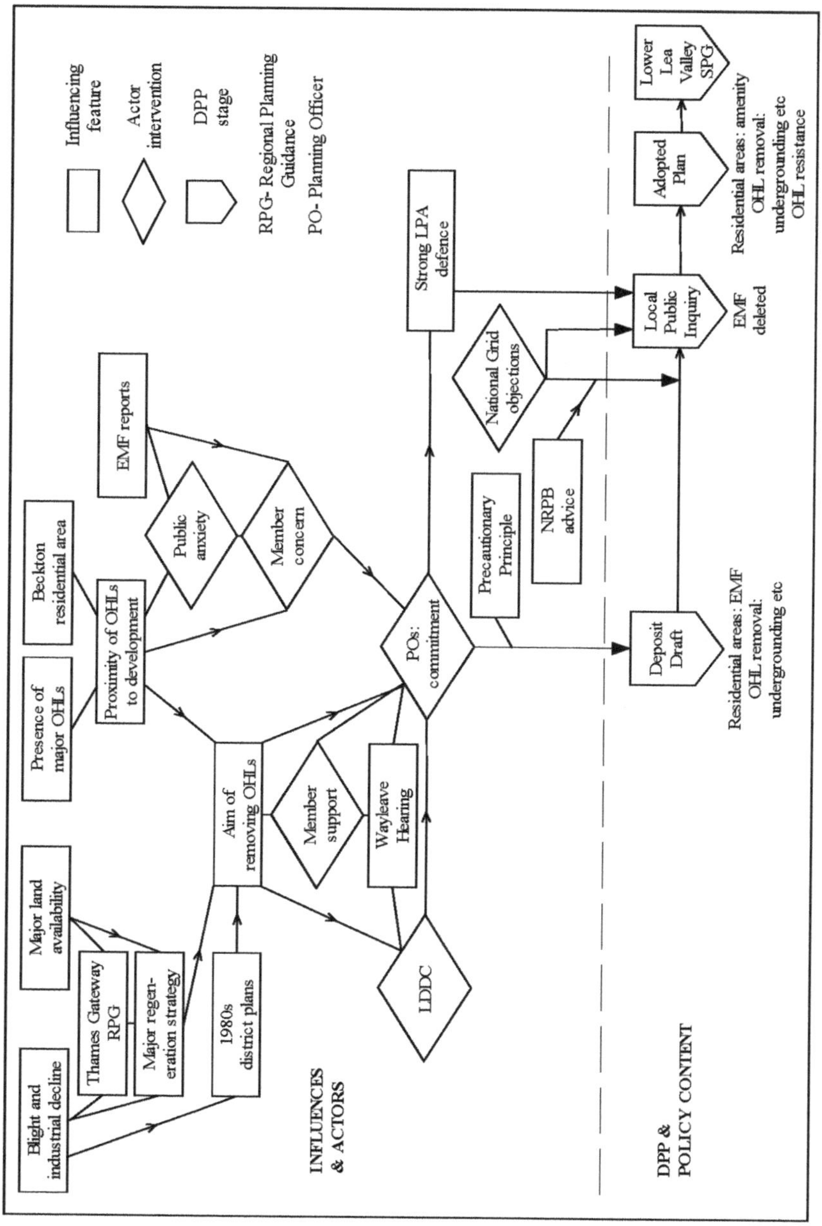

Figure 6.13 The process of policy formation by Newham Council

CHAPTER 7

Case Studies (2): Rural Areas

7.1 INTRODUCTION

The case studies in this chapter explore the concerns of local planning authorities (LPAs) in large, predominantly rural areas. Tynedale Council covers a northern upland and valley area; Norfolk County Council covers a lowland, mostly rural county, and Snowdonia National Park Authority covers a diverse mountain and moorland area. In addition to including Snowdonia's designation as a national park, these authorities cover a wide range of areas and sites that have been designated for their landscape, scientific or historic interest. The three authorities concentrate primarily on the possibility of new HVDT installations in their areas and seek particularly to defend designated areas and sites from their intrusion. One feature in common between these case studies is some degree of controversy over existing installations.

TYNEDALE COUNCIL

7.2 INTRODUCTION

Tynedale Council (TC) is a district council in Northumberland in the north of England. It covers a large, mostly rural upland, area and is bisected by the Tyne Valley. The valley contains the main settlements, including the town of Hexham, while the countryside covers parts of the Northumberland National Park, the North Pennines Area of Outstanding National Beauty (AONB), the World Heritage Site of Hadrian's Wall and includes other designated areas. The authority has recently adopted its first district-wide local plan (TC 2000); this case study was undertaken during the period following adoption.

The plan places immediate emphasis on the quality of Tynedale's environment, stating as one of its underlying themes "the need to protect and enhance the natural and built environment of the district, maintaining and creating an environment which makes Tynedale an attractive and pleasant place to live, work and visit" (TC 1996). Encouraging tourism is an explicit aim of this strategy; the plan refers to the decline in traditional rural industries and looks to the development of tourism as a means of improving the local economy. The district is described as containing a wealth of natural and man-made features that have the potential to draw visitors.

7.3 ARTICULATION OF POLICY

A policy on high-voltage overhead lines (OHLs) has been brought through plan preparation under the section "Community Services and Infrastructure". It is concerned with the location of new OHLs and resists their siting within areas valued for their natural or man-made features and near residential and recreational areas. (An accompanying policy deals in similar terms with OHLs of less than 132 kV.) Figure 7.1 shows key elements of the policy. It lists those areas and sites for which protection is sought; this includes areas designated for their landscape or nature conservation interest, both statutorily, such as the AONB, and locally, such as the Area of High Landscape Value. (The Northumberland National Park is not referred to, as it is not included in the plan area.)

The reason given for opposition to new OHLs in designated areas is that they "can be visually prominent and damaging to the landscape" (TC 2000). The visual impact of OHLs is also stated as justification for OHLs to avoid residential areas. The early drafts of the plan (TC 1994, 1996) also cite possible health and safety effects of OHLs as reasons for routeing lines away from both residential and recreational areas. According to these early versions of the policy, the council would only drop its opposition to a new OHL in any of the specified areas in the case of "overriding national need" for the development and if there is no practical alternative. Finally, the policy encourages the undergrounding of lines if they must pass through sensitive locations.

Support for the OHL policy was received from English Nature and the Royal Society for the Protection of Birds. However, objections to various aspects of the policy were received from the electricity industry (NG and Northern Electric). Their objections to the implied health and safety risks in residential and recreational areas were accepted by the council before the plan inquiry; the justification for resisting OHLs in these areas was changed to "amenity" reasons. Objections to the call for undergrounding some lines were also accepted; the policy was reworded to acknowledge the difficulties involved in undergrounding, and to recognise the role to be played in minimising visual impact by careful OHL routeing. This change was also in response to a comment from English Heritage, concerned about the possible harm to archaeological remains from placing lines underground.

1996

Tynedale District Local Plan Deposit Draft

- Overhead line policy

↓

2000

Tynedale District Local Plan Adopted Version

- Overhead line policy

Overhead Electricity Power Lines

...the impact of larger electricity supply lines (132kv or over)... can be visually prominent and damaging to the landscape...
The visual impact and concern over the possible affects upon the health and well-being of residents living close to lines makes it desirable to route lines away from residential areas. Whilst the National Grid Company undertakes safety measures, it is felt that location close to recreation areas should be avoided.

...Tynedale District Council will normally oppose the siting of overhead power lines of 132kv or over... in the following locations unless there is an overriding national need for the development and the cost and scope for the development has been examined and no alternative sites are available in less sensitive areas:
(a) within or in prominent visual locations adjacent to the North Pennines Area of Outstanding Natural Beauty;
(b) in prominent visual locations within the Area of High Landscape Value;
(c) in areas of nature conservation interest;
(d) in Conservation Areas or in the vicinity of sites of particular historic, archaeological or architectural interest;
(e) close to existing or proposed residential development;
(f) areas used for informal or formal recreation including playing fields, public open spaces and country parks.

In sensitive locations Tynedale District Council will seek the location of power lines underground, unless this would cause greater damage to the ecology.

Figure 7.1 Policy development by Tynedale Council

More contentious debate was held, including at the public local inquiry, in relation to the list of areas where OHLs would be resisted, and the associated criteria under which the council would not oppose OHLs. The electricity industry argued that

- the range of areas listed was such that a resistance to OHLs would be established across most of the plan area;
- areas carrying non-statutory designations, such as Areas of High Landscape Value, should not receive the same standard of protection as nationally designated areas;
- nature conservation sites and recreational areas do not necessarily need protection from OHLs;
- the council's criteria for not opposing OHLs (overriding national need and lack of alternative sites) are impractical; for instance, it may not be possible to demonstrate national need for local distribution lines.

The council justified its criteria with reference to national planning guidance (*PPG 7*: DoE 1992c, 1997b) and the Northumberland Structure Plan (Northumberland County Council 1996), from which the wording of the criteria was directly drawn. These documents specify the conditions under which major developments (specifically including electricity transmission lines in the case of the structure plan) might exceptionally be allowed in AONBs. However, the council did accept that these criteria should not apply to areas that do not carry national designation.

The inquiry inspector supported the council's approach in seeking to protect the areas listed, including nature conservation sites and recreational areas. However, he recommended that a distinction should be drawn between the level of protection sought for nationally designated and other areas. The adopted policy reflects this recommendation, in that the original criteria for not opposing OHLs are only applied to nationally designated sites, while less strict criteria apply to the other areas listed.

7.4 KEY POLICY ISSUES

7.4.1 Awareness of HVDT installations

The plan explicitly refers to both actual and potential HVDT installations in the district. The OHL policy mentions the NG-owned 275 kV line that runs across the entire district through the Tyne Valley (TC 2000). It also states that NG has no plans to construct new OHLs in the area, suggesting that the council remains attentive to the possibility of new lines in the future. Moreover, the deposit draft of the plan refers to the likely need for a new primary substation in the district to improve local electricity supply (TC 1996).

A planning application for a 275 kV substation was in fact made in 2000 by NG. The area is within the Area of High Landscape Value and the "landscape setting" of the Hadrian's Wall World Heritage Site (see the following text), has wildlife and archaeological value and is within the Green Belt; the sites considered for the substation are also close to the small settlement of Fourstones. The application has caused considerable opposition from local residents and amenity groups concerned about a number of issues: visual impact, perceived health risk, noise and impingement on designated areas and monuments (TC Planning Officers A and B 2001).

Although this controversy post-dates the main preparation of the plan, the anticipation of the substation in the deposit draft indicates that there has been longer-standing sensitivity to HVDT installations as a local planning issue. The presence of both existing and planned installations in the district therefore appears to have contributed to bringing OHLs to the fore as a policy topic.

7.4.2 Protection of valued natural and man-made features

The main focus of the OHL policy is on the protection of areas and sites of natural and man-made interest. As inferred above, the locations specified in the policy cover a significant proportion of the plan area:

- The North Pennines AONB covers a large area to the south of the Tyne Valley.
- The Area of High Landscape Value covers much of the district not otherwise designated; "prominent visual locations" within this area are covered by the policy.
- Areas of nature conservation interest: the plan area contains a large number of sites of varying international, national and local importance.
- Conservation Areas and historic, archaeological and architectural sites; the World Heritage Site of Hadrian's Wall, running the entire width of the District, is the most prominent of these. Furthermore, the plan defines a corridor approximately 4 km wide along the length of the Wall as its "landscape setting", subject to strict development control.

This unusual concentration of valued areas and sites leads the council to be "very landscape and nature conservation oriented" (TC Planning Officer A 2001); they are seen as combining to produce a unique landscape deserving special protection. This is illustrated by the landscape assessment of the district carried out by the council as a means of justifying the Area of High Landscape Value: "...landscapes of a particularly high value which arise as a result of both the quality of the landscape and its visual, cultural, social and historic characteristics" (TC 1998a). Similarly, in relation to the landscape setting of Hadrian's Wall, the plan seeks to protect not just the corridor defined on the proposal's map, but the entire zone within which the Wall is visible, including those from distant hills (TC 1996). This overall stance is summarised in the plan's environmental strategy, which states the need "to conserve the district's natural and historic assets" (TC 1996). The council has based this approach on national, regional and structure plan policy encouraging protection of designated areas (DoE 1993a, 1994, Northumberland County Council 1996).

It is in this context that the council's attitude to OHLs should be seen. The landscape assessment referred to above specifically draws attention to the effect of OHLs, describing them as negative elements on the landscape "considered to have a downgrading effect" (TC 1998a). This parallels the description of OHLs in the plan as "visually prominent and damaging to the landscape" (TC 2000). This opinion appears to have been most strongly expressed at officer level, as, for example, at inquiry stage, where OHLs were called *intrusive elements*, whose "pervasive nature reduces landscape quality generally" (TC 1998b). The presence of existing OHLs contributes to this perspective; Figure 7.2 shows an OHL in close proximity to the World Heritage Site of Hadrian's Wall.

Figure 7.2 An OHL within the landscape setting of Hadrian's Wall, Tynedale

The effect of lower-voltage overhead wires in a designated area was raised in the context of a local planning appeal in 1997. In this case, the council had granted planning permission for a farm conversion in the AONB with the condition that no services should be provided on overhead wires in the interests of the appearance of the area. The appeal inspector agreed with the council on this matter. This decision was quoted by the council at the inquiry into the plan in defence of another policy on low-voltage OHLs (TC 1998b). Although not directly relevant to high-voltage lines, this case indicates that electricity equipment has been a recent practical planning issue; the case of the Fourstones substation referred to above also illustrates this point.

The council's preservationist approach to the district's "natural and historic assets" is based partly on the aim of encouraging tourism. The plan sees the growth of tourism as a means of boosting the local economy and countering the decline of traditional activities, such as agriculture, forestry and quarrying. "The superb natural environment and a plethora of attractive and historic man-made features ... provide an excellent base from which to develop tourism in the District" (TC 1996). Protection of these features is therefore regarded as an economic priority, and policies that aim at conserving the setting of Hadrian's Wall, for example, have tourism potential partly in mind. The OHL policy is consistent with and contributes to this strategy: "We use the landscape to help the economy of the district... the [OHL] policy is to protect the landscape in order to encourage tourism" (TC Planning Officer A 2001).

Finally, the council was able to turn to its broader policy framework in the preparation of this aspect of its policy. The preparation of the structure plan preceded the council's plan by some months and provided the wording for the conditions incorporated into the early versions of the OHL policy (TC Planning Officer A 2001); the council was also able to rely upon both the structure plan and national planning guidance in the defence of this policy.

7.4.3 Protection of residential and recreational areas

The OHL policy originally aimed at protecting residential and recreational areas mainly from the perceived health and safety risks of OHLs. This aspect of the policy appears to have been primarily a concern of council officers. When challenged by the electricity industry on this point, they were prepared to modify the policy accordingly; this was in the knowledge of difficulties faced by other authorities who had attempted to take a similar approach and because "we didn't have the necessary skills/knowledge to argue such a case at inquiry" (TC Planning Officer A 2001).

There do not appear to have been strong public attitudes in the district about possible health and safety effects of OHLs, except in the case of the Fourstones substation. However, there have been increasing anxieties about the perceived health risk of electromagnetic fields (EMF) from telecommunications equipment, both among the public and elected members; this has been in response to a number of new installations (TC Elected Member 2001). Moreover, the deposit draft of the plan includes a policy on telecommunications equipment expressing concerns similar to that on OHLs, including visual and heath effects (TC 1996). Although the plan does not link the policies on OHLs and telecommunications equipment, it is conceivable that some mutual reinforcement of parallel concerns has taken place.

7.5 CONCLUSION

The predominant concern of TC's first district-wide local plan is for the preservation of the area's unusual concentration of natural, archaeological and historic assets; this is partly with a view towards the contribution that these assets can make to the growth of tourism. These aspects of the environment are therefore attributed value in the plan and given a considerable degree of policy attention. It is within this context that the OHL policy's aim of protecting valued sites and areas has been advanced. The following features have been particularly instrumental:

- the presence of existing OHLs has raised the question of their place within the landscape;
- OHLs have been regarded as having detrimental effects at a professional planning level, including through a landscape assessment of the district;
- national, regional and structure plan policy has provided a framework for the protection of designated areas from OHLs;
- the possible effects of HVDT installations upon landscape, and so on, have recently been brought into the local planning arena by the application for the Fourstones substation (and, to some degree, by the planning appeal in the AONB).

The OHL policy's aim of protecting residential and recreational areas was initially concerned with perceived health and safety risks; in addition to arising as an officer initiative, this parallels anxieties about telecommunications equipment. The shift to "amenity" reasons for protecting these areas brings this aspect of the policy closer to the policy's other aim of protecting valued areas. The components of policy formation by TC are depicted in Table 7.1. The inter-relation of the principal components and the progression of policy through the development plan process (DPP) are summarised in Figure 7.3.

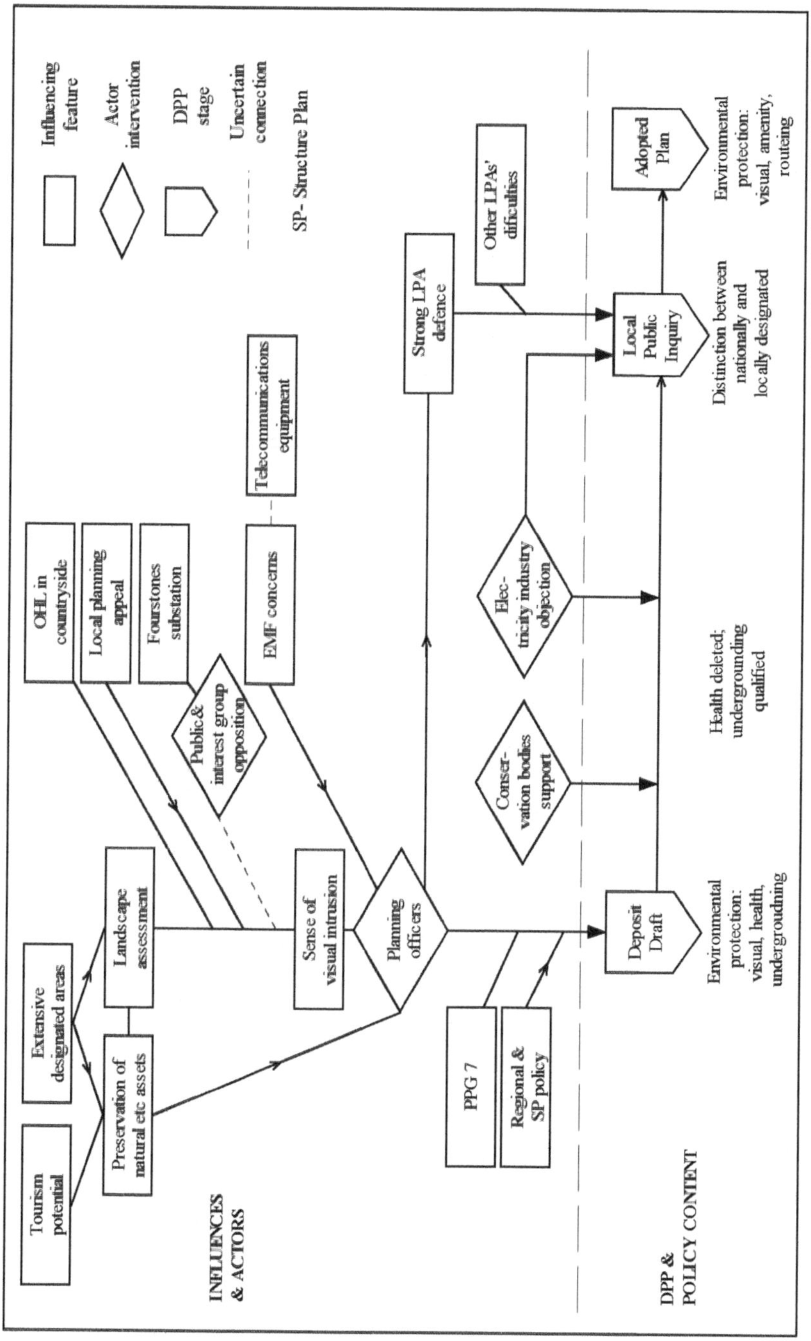

Figure 7.3 The process of policy formation by Tynedale Council

Table 7.1 Components of policy formation by Tynedale Council

BROAD CATEGORIES		WITHIN-CASE COMPONENTS	CROSS-CASE COMPONENTS*
INFLUENCING FEATURES	GEOGRAPHICAL, ECONOMIC AND SOCIAL FEATURES	Extensive designated areas and sites	Extensive designated areas and sites
		Extensive rural areas	Extensive rural areas
		Contribution of natural, etc. features to growth of tourism	Potential for tourism
		OHL traversing district	Prominent installations in countryside
	HVDT INSTALLATIONS (AND TELECOMMUNICATIONS)	Installations regarded as damaging to landscape areas, etc.	Installations affecting designated areas/sites
		Fourstones substation	Recent/current HVDT projects
		Public anxieties	Local opposition to telecommunications equipment
	POLICY FRAMEWORK	PPG7: major developments	Supportive national planning policy
		Supportive regional guidance and structure plan policy	Supportive regional/structure planning policy
	PLANNING EXPERIENCE	Fourstones substation	Recent/current HVDT projects
		Local planning appeal	Electricity installation development control cases
		Knowledge of other authorities' difficulties regarding EMF	Knowledge of other LPAs' experience
	PLAN THEMES	Preservation of natural, historical assets	Protection of environmental features
ACTORS	PLANNING OFFICERS	Policy initiated by officers	Strongly supportive of policy
		Local landscape assessment and designations	Landscape valuation/assessment
		EMF concerns	Study of EMF research
	INTEREST GROUPS	Opposition to Fourstones substation	Opposition to HVDT projects
	PUBLIC		Opposition to HVDT projects
	ENVIRONMENTAL BODIES	Supporting proposed policy	Supporting aspects of policy
	ELECTRICITY INDUSTRY	Opposing proposed policy	Opposing aspects of policy
DPP	REPRESENTATIONS	Conservation bodies supporting proposed policy	Supporting HVDT-related policy
		Electricity industry opposing proposed policy	Opposing HVDT-related policy
	LPA DEFENCE	Strongly defending policy against representations	Strongly defending policy against objections
	Local public inquiry (LPI) / Examination in public (EiP)	Contestation	Contestation
		Inspector recommended alterations	Inspector/Panel recommended alterations
	BROAD THEMES	Protection of designated areas/sites	Countryside protection
		Protection of residential /recreational areas	Residential area protection

BROAD CATEGORIES		WITHIN-CASE COMPONENTS	CROSS-CASE COMPONENTS*
POLICY CONTENT	EFFECTS OF HVDT INSTALLATIONS ON:	Visual quality of landscape	Visual quality: countryside
		Visual quality of residential areas	Visual quality: urban areas
		Nationally and locally designated areas	Designated areas: national
			Designated areas: local
		Nature conservation areas	Ecological interests
		Archaeological sites	Archaeological sites
		Historical sites	Historical sites
		Health	Public health (EMF)
		Safety	Public safety
		Residential areas	Residential, etc. areas
		Recreational areas	Recreational areas
	MEASURES TO OFFSET EFFECTS	Opposition to new OHLs	Opposition to new HVDT installations
		Routeing OHLs away from residential areas	Careful routeing of new OHLs
		Undergrounding lines in sensitive areas	Undergrounding of lines
		Routeing OHLs away from residential areas	Distancing new OHLs from existing development

* These refer to components defined for the case studies as a whole; see Section 9.2.1.

NORFOLK COUNTY COUNCIL

7.6 INTRODUCTION

Norfolk County Council (NCC) is a county authority in the east of England. Norfolk is a large, mostly rural county containing diverse forms of lowland landscape and habitat and many areas designated for their landscape or wildlife importance, including the Broads. The largest settlements are Norwich and the towns of King's Lynn and Great Yarmouth.

The council adopted a replacement structure plan in 1999 (NCC 1999). This is based upon the overall aim of sustainable development, in which emphasis is placed upon the quality of Norfolk's natural and built environment. This is seen as a draw to inward investment and a means of offsetting some of the county's economic and social difficulties. High priority is thus given to the conservation and enhancement of the environment; this is also seen as a contribution to wider biodiversity.

Administratively, the county is made up of seven districts. Each district authority, along with the Broads Authority, has an adopted local plan that is currently being revised to be in conformity with the present structure plan.

7.7 ARTICULATION OF POLICY

NCC's plan does not treat HVDT installations as a separate topic, but makes reference to them in two policies that deal with proposals for the generation of energy. These are concerned with the location and effects of both conventionally fuelled power stations (non-renewable energy) and renewable energy projects; these are seen as sufficiently strategic issues to be treated at structure plan level. The policies seek to prevent adverse environmental impacts arising from these developments, especially in areas with landscape, conservation, recreational, and so on, values; in the case of renewable energy schemes, protection is particularly sought for designated areas. The possible effects of energy generation projects upon sensitive areas include any impacts resulting from power lines that are required to connect new generating installations to the electricity network. These policies state that any new power lines associated with energy projects should be placed underground (the same applies to pipelines). They also indicate that the location of new schemes relative to the existing transmission network should be taken into account. These policies were introduced at the early consultation stage of the plan (NCC 1997) and carried through to adoption with substantially similar wording. They are summarised in Figure 7.4.

1998
Norfolk Structure Plan
Deposit Version

- Renewable and non-renewable energy policies

1999
Norfolk Structure Plan
Adopted Version

- Renewable and non-renewable energy policies

Non-Renewable Energy

Proposals for the development of conventionally fuelled power-stations will only be acceptable where:

- **there are no serious environmental or transport impacts; and**
- **all options to route power lines... linking a power plant to the electricity network... underground have been investigated and, where appropriate, incorporated in the proposal.**

Renewable Energy

Renewable energy developments will be encouraged to locate where their scale, siting or cumulative effect would not have a significant adverse environmental impact... Careful scrutiny will be given to the scale, type and impact of proposals for renewable energy development in or adjacent to the Broads, the AONBs, the Heritage Coast, SSSIs, Special Protection Areas (birds), Special Areas of Conservation, NNRs, Ramsar sties and LNRs...

...proposals for conventionally fuelled power stations and renewable energy schemes will need to address the following issues:

- impact (including cumulative effects) on landscape, nature conservation and other environmental considerations...
- conservation areas and settlements...
- sites of archaeological importance...
- communication networks...
- amenity of the surrounding area
- location in relation to the transmission network where the development is to be linked into the electricity network and the need to route associated power lines and pipelines underground
- that the power plant will not lead to the unsustainable transportation of resources over long distances...
- areas of recreational or tourist value or potential...
- agriculture.

Figure 7.4 Policy development by Norfolk County Council

During the deposit stage of the plan, representations were made on the policies by the electricity industry. Eastern Generation Ltd objected to the criteria by which the council would oppose new generating projects and to the stance taken on undergrounding associated power lines. NG also objected to the call for undergrounding, on the grounds of technical and environmental difficulties and cost, stating that careful line routeing was the preferred means of minimising environmental damage. The British Wind Energy Association, along with other groups, argued for a more positive approach to wind energy. Similarly, the Government Office for the Eastern Region and two of the county's district authorities suggested a more flexible approach towards renewable energy proposals.

The policies were debated at the Examination in Public held into the plan (to which Eastern Generation Ltd was invited as a participant). Discussion focused on the likely increase in applications for projects in response to government targets for renewable energy. Eastern Generation Ltd argued that greater account should be taken of the county's renewable energy potential, while the council maintained that priority should be given to the protection of Norfolk's countryside from, for example, the visual intrusion of wind turbines. The Panel generally supported the council's position, but recommended a less restrictive approach to renewable energy projects. On the question of power lines associated with new generating schemes, the Panel supported "the broad contention that power lines should be routed underground where possible and/or justified" (NCC 1998), though recommended a less stringent wording in the policy. The council accepted these recommendations and modified the wording of policy accordingly; for example, the call to make "every effort" to route OHLs underground was changed to one of investigating the options for undergrounding (NCC 1999).

7.8 KEY POLICY ISSUES

7.8.1 Effects of energy generation projects

The energy generation policies seek to ensure environmental protection in the face of likely proposals for renewable and non-renewable energy projects; powerlines are seen as possible associated developments.

The county council, along with the relevant districts, has dealt with a number of energy proposals since the mid-1990s. There have been applications relating to two major gas-fired power stations; these (along with a similar project just into Lincolnshire) sought to take advantage of North Sea gas pipelines terminating on the East Anglian coast and government policy favouring gas as a fossil fuel (the "dash for gas"). There has also been a run of proposals for smaller-scale renewable energy schemes, especially wind turbine projects and biomass-fuelled plants. Furthermore, the council expects a continuation of applications, especially in relation to renewable energy sources. This expectation reflects government targets for renewable energy generation, quoted in the plan as 10% of national electricity supply by 2010. Also, the plan refers to the report of the Eastern Region Renewable

Energy Planning Study, which assesses the potential of the region for renewable energy resources, partly in order to inform structure plan reviews (Terence O'Rourke 1997). Possible renewable sources are outlined in the plan. The policies therefore represent an awareness of the momentum and changing character of energy schemes in the county and of the potential environmental issues arising from them:

> ...we'd actually adopted an interim renewable energy policy, with quite a few schemes coming forward in the 90s, it was decided that we would have to develop policies to cover all the eventualities ... (NCC Planning Officer 2001).

The formation of policy has also been precipitated by the fact that some of the recent energy projects have given rise to particular environmental concerns; powerlines have figured prominently in this respect. These cases can be summarised as follows (all cases: NCC 1996):

1. The most controversial case related to the gas-fired power station at King's Lynn; the proposed upgrading of this plant led to an application for a new 400 kV OHL to connect the power station to the grid. The proposed route for the line crosses low-lying Fenland, through a locally designated landscape area, for 3 km. County council officers made a recommendation that no objection should be made to the line, but this was overturned at committee stage by elected members, who argued that the line should be placed underground in order to prevent visual intrusion to the landscape. A similar position was taken by members for the respective district during their discussion of the case.
2. The construction of a new gas-fired power station at Great Yarmouth required the upgrading of the Great Yarmouth – Norwich 132 kV OHL (30 km long), sections of which cross an urban area and the Broads. The proposals included undergrounding for these sections of the upgraded line; the county council did not object to these proposals, but concern was expressed by members and a local parish council that more extensive undergrounding should be carried out.
3. An application for a combined heat and power plant, fuelled by gas derived from sugar beet, required the construction of a lower-voltage (33 kV) OHL across a locally designated landscape area. No objection was raised, but the undergrounding of one section of the line was recommended by the council.

These cases placed the environmental effects of energy generation, especially the associated need for new or upgraded OHLs, firmly within the local planning arena, precisely at the time when review of the structure plan was getting underway. The strength of feeling aroused by the OHLs required by the gas-fired power stations has also been illustrated more recently: they have been referred to while the council has been considering its support for a parliamentary Early Day Motion on "Power Lines and the Environment" (NCC Landscape Architect 2001).

7.8.2 Protection of sensitive areas

Insofar as powerlines are concerned, the council's energy generation policies aim to ensure the protection of landscape and wildlife, especially in designated areas. An interest in recreation and tourism and in "amenity" is also implied. The policies are therefore concerned with the effect of powerlines on the quality of sensitive areas, arising especially from the visual impact of OHLs. The cases referred to above demonstrate the same interests. For example, the proposed upgrading of the Great Yarmouth – Norwich line across the Broads led to calls that more of the line should be placed underground because of its various visual effects:

> I'm very concerned about the environmental aspect, particularly if you understand that Norfolk, it's not all low lying, but part of my area actually includes the Broads and it is therefore relatively flat... I was arguing that the lines should not be altered in such a way as they damage the environment of historic or listed buildings or some ancient settlement, and that the existing lines should be undergrounded near the modern housing development (NCC Elected Member 2001).

This case showed a strong interest in landscape; even though undergrounding was planned for the part of the line that crossed the Broads, there was a feeling that adjacent areas equally merited protection because of their landscape importance (NCC Elected Member 2001). This typifies the value placed more widely upon the county's landscape and a perception that OHLs are an intrusive element within it, as evidenced by this officer opinion:

> The lines are ... potentially more dominant over a higher land area, because you do get the long views, and the long views over open countryside are something which are particularly valued locally. So I think there's a ... member and also general public value put on ... expansive views over an open landscape ... to the people living locally... its something they've grown up with ... then if you suddenly start putting 400 kV lines marching in all directions over it, it obviously does impinge on it (NCC Landscape Architect 2001).

The open landscape of Norfolk is therefore perceived as making it more vulnerable to the visual effects of OHLs. Moreover, a number of existing OHLs are considered to be particularly intrusive:

> ... when they built the southern [Norwich] bypass, it suddenly opened up all the views to all the powerlines ... I think I did a count once ... there were something like 13 different styles of pylons ... there must be a complete collection. That was something that was a bit shocking to people (NCC Landscape Architect 2001).

The area with the highest concentration of OHLs is around the Walpole 400 kV substation, in the west of the county; nine OHLs converge on the substation, crossing low-lying Fenland (Figure 7.5). The proposed OHL from the King's

Lynn power station would be in this vicinity, accounting for some of the opposition: "... they've got so many there ... yet another large line is another nail in the coffin, that area between King's Lynn and Sutton Bridge, it's really been wrecked by overhead lines" (NCC Landscape Architect 2001).

Figure 7.5 OHLs in the vicinity of the Walpole 400 kV substation, Norfolk

The desire to prevent further OHLs from having a detrimental effect on landscape is consistent with the plan's wider emphasis on landscape and environmental protection. This theme is given prominence throughout the plan as an element of the sustainable development strategy that the council sets itself:

> The County Strategy is to provide for sustainable development in Norfolk by (i) giving high priority to protecting and enhancing the wealth, variety and distinctiveness of Norfolk's natural and built environment ... (NCC 1999, p. 11).

This emphasis is partly a reflection of the number and extent of designated areas and sites in the county (NCC Planning Officer 2001). A range of policies aims to preserve the landscape, wildlife and heritage features of the county and also to improve the environment where it has been damaged by past actions. For example:

> In Norfolk, it is important to protect the countryside outside areas of national designation as unsympathetic development will gradually erode the distinctive character of the county as a whole (NCC 1999, p. 26).

These policies are seen as applying widely to development proposals, potentially including matters such as energy generation (NCC Planning Officer 2001).

This emphasis of the plan can be located within a longer history of environmental concern of the council: "... it's an iterative thing over time where there's been a commitment by both members and by officers to protecting the environment of Norfolk" (NCC Planning Officer 2001); "... members in particular are very committed to protecting the rural environment" (NCC Landscape Architect 2001).

The plan suggests the undergrounding of powerlines as a means of alleviating their effects on sensitive areas. The possibility of placing lines underground has been presented by the actual undergrounding of sections of the Great Yarmouth – Norwich line (Figure 7.6) and also of a number of lower-voltage supply lines in rural areas. Some of the lower-voltage lines were placed underground as a negotiated compensatory measure for the upgrading of the Great Yarmouth – Norwich line, in the area where the greatest opposition to the upgrade occurred (NCC Elected Member 2001). There have therefore been precedents in the county for the undergrounding of lines, albeit not wholly for high-voltage lines. The greatly differing technical and cost implications of placing lower- and higher-voltage lines underground have been recognised, particularly at officer level. However, members have remained insistent upon their preference for this measure, as evidenced by their overturning of the officer recommendation on the King's Lynn OHL case (NCC Landscape Architect 2001, NCC Elected Member 2001).

Figure 7.6 Sealing-end compound at the start of an underground cable crossing the Norfolk Broads

A desire for the undergrounding of lines has also been stated in four of the local plans prepared within Norfolk (Breckland Council 1999, Broads Authority 1997, Great Yarmouth Borough Council 2000, North Norfolk District Council 1998). These, along with another of the county's local plans (South Norfolk Council 1997), variously express concern for the visual effects of lower- and higher-voltage OHLs. So there has been a consensus at both tiers of local government within Norfolk for minimising the visual impact of OHLs, the undergrounding of lines being seen as a favoured option.

The plan also implies that the impact of OHLs can be reduced by locating new energy schemes close to the existing electricity network. This suggestion arose in relation to a proposed wind turbine, where the question of its connection to the grid was discussed (NCC Planning Officer 2001). Here, too, the experience of a specific project has informed the ensuing formation of policy.

7.8.3 EMF-related health concerns

At no stage of its preparation did the plan make any reference to the alleged health effects of EMF. However, it appears that underlying anxieties about possible EMF-related health risks have contributed to opposition to OHLs. Public concerns on the subject were evident in the early to mid-1990s (NCC Landscape Architect 2001) and were expressly voiced in relation to the upgrading of the Great Yarmouth – Norwich line, which passes close to village housing: "I just think on a precautionary basis they should be further from houses" (NCC Elected Member 2001).

7.9 CONCLUSION

NCC's concerns with HVDT installations, specifically OHLs, are set within the context of policies designed to limit the environmental effects of energy generation projects. These policies are in anticipation of proposals for new energy schemes in the county, following on from recent applications. This trend demonstrates the favourable position in which Norfolk finds itself for certain types of generation, increasingly from some renewable sources and reflects government energy policy. OHLs associated with these schemes have emerged as one of the principal planning concerns relating to energy generation, possibly more contentious than the generation projects themselves. OHLs are seen as detrimental to the landscape of the county as a whole and to the quality of its designated areas and sites. The undergrounding of lines is put forward as the preferred mitigation measure.

A number of features have contributed to this policy approach:

- the council has set itself a sustainable development strategy within which the protection of environmental assets, such as landscape, is given a high priority; this is consistent with the council's history of environmental concern, and a reflection of the large number of designated areas;
- council officers judge Norfolk's landscape to have considerable local value;

- existing OHLs are visually dominant in several locations in the county; these are perceived as an intrusive element in the mostly open landscape;
- recent energy-related proposals have raised the prospect of further OHLs being constructed, including those in sensitive areas;
- the experience of considering recent proposals has provided officers and members with policy notions; similarly, a precedent has been set for undergrounding OHLs in certain areas;
- members have pursued the OHL issue with some vigour, on one occasion against the advice of officers;
- district councils have added weight to the perceived need to minimise the impact of OHLs and to the idea of undergrounding.

In addition, a concern for the alleged health effects of EMF has been voiced, though it has not found policy expression. The components of policy formation by NCC are depicted in Table 7.2. The inter-relation of the principal components and the progression of policy through the DPP are summarised in Figure 7.7.

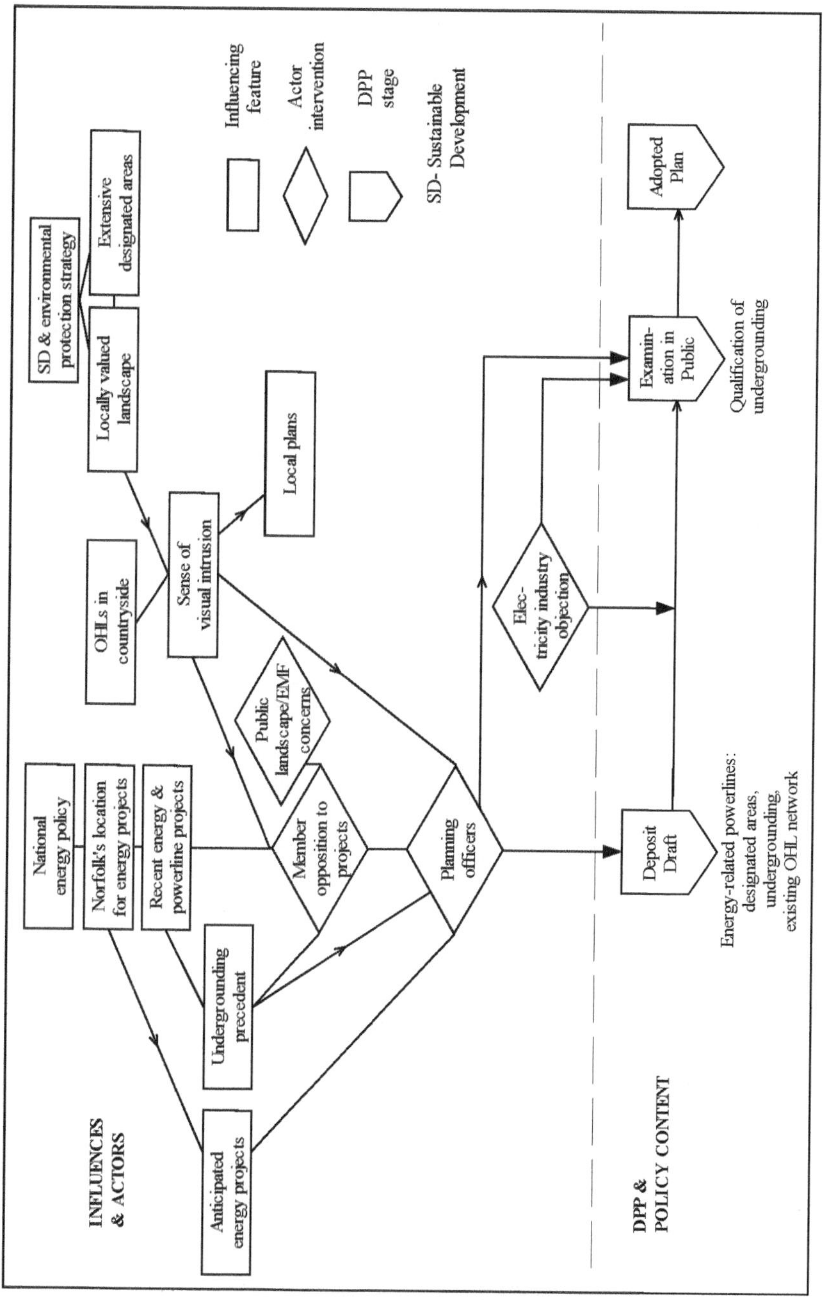

Figure 7.7 The process of policy formation by Norfolk County Council

Table 7.2 Components of policy formation by Norfolk County Council

BROAD CATEGORIES		WITHIN-CASE COMPONENTS	CROSS-CASE COMPONENTS*
INFLUENCING FEATURES	GEOGRAPHICAL, ECONOMIC AND SOCIAL FEATURES	National energy policy	Pressure for energy (especially renewable) projects
		Extensive designated areas and sites	Extensive designated areas and sites
		Extensive rural areas	Extensive rural areas
		Favourable location for energy projects	Potential for energy generation projects
	HVDT INSTALLATIONS (AND TELECOMMUNICATIONS)	OHLs visually dominant and taken to be detrimental to designated areas	Prominent installations in countryside
			Installations affecting designated areas/sites
		Recent energy and associated powerline projects	Recent/current HVDT projects
		New lines expected for energy projects	Perceived likelihood of future HVDT projects
	POLICY FRAMEWORK	Local plan policies resisting OHLs	Supportive local planning instruments
	PLANNING EXPERIENCE	Recent energy-related powerlines	Recent/current HVDT projects
		Undergrounding precedents	Local precedents of measures to offset effects of HVDT
	PLAN THEMES	Sustainable development strategy	Sustainability
		Protection of environmental assets	Protection of environmental features
ACTORS	PLANNING OFFICERS	Landscape taken to have considerable local value	Landscape valuation/assessment
	ELECTED MEMBERS	Insistence on aspects of policy	Strongly supportive of policy
		Opposition to OHL projects	Opposition to HVDT projects
		Landscape and designated area protection concerns	Landscape, etc. protection concerns
		Voicing of EMF health concerns	EMF health/amenity concerns
	PUBLIC	Landscape and designated area protection concerns	Landscape, etc. protection concerns
		EMF health concerns	EMF health/amenity concerns
	ELECTRICITY INDUSTRY	Opposing proposed policy	Opposing aspects of policy
DPP	REPRESENTATIONS	Electricity industry and others opposing proposed policy	Opposing HVDT-related policy
	Local public inquiry (LPI)/Examination in public (EiP)	Panel recommended alterations	Inspector/Panel recommended alterations
	BROAD THEMES	Environmental effects of energy generation	Countryside protection
	EFFECTS OF HVDT INSTALLATIONS ON:	Landscape	Visual quality: countryside
		Nationally and locally designated areas and sites	Designated areas: national
			Designated areas: local
		Conservation areas	Ecological interests
		Archaeological sites	Archaeological sites
		Amenity	Amenity

BROAD CATEGORIES		WITHIN-CASE COMPONENTS	CROSS-CASE COMPONENTS*
		Areas of recreational and tourism value	Recreational areas
	MEASURES TO OFFSET EFFECTS	Undergrounding of lines	Undergrounding of lines
		Location of energy schemes near existing transmission network	Siting new energy schemes near existing OHLs

* These refer to components defined for the case studies as a whole; see Section 9.2.1.

SNOWDONIA NATIONAL PARK AUTHORITY

7.10 INTRODUCTION

Snowdonia National Park Authority (SNPA) is the local planning authority for the Snowdonia (*Eryri* in Welsh) National Park in North Wales. The park covers extensive areas of mountain and moorland, along with coastal stretches, river valleys, lakes and woodland. Many areas and sites carry additional wildlife, geological, archaeological or historic designation, some of international status. Snowdonia contains a number of small towns and villages and is an important focus of Welsh language and culture.

The National Park Authority (NPA) prepared its first park-wide plan (the Eryri Local Plan) under the Gwynedd Structure Plan; it was adopted in 1999. However, the NPA took on unitary authority status following local government reorganisation in 1996 and is now responsible for the preparation of a unitary development plan; this process began in 2000. References to the "plan" given in the following paragraph are to the local plan.

The plan takes as its starting point the statutory purposes of national parks; it places emphasis on the conservation of Snowdonia's natural and cultural features, appropriate leisure activities in the park and the well-being of local communities. The plan allows for change to take place within the park, but sees this in terms of sustainability so that new development and activities are protective of the environment and the park's character; Snowdonia is considered to be particularly unsuitable for the location of major developments.

7.11 ARTICULATION OF POLICY

The Eryri Local Plan (SNPA 1999) makes a number of references to "transmission lines/systems", mostly in the wider context of energy generation (Figure 7.8).

1977
Snowdonia
National Park Plan

- Major civil engineering works

→

1986
Snowdonia
National Park Plan Review

- Electricity development policies

The National Park Committee will prepare a list of the most damaging examples of unsightly pylons... with a view to undergrounding or adjustment of the lines...

The National Park Authority "will oppose the further violation of the great natural beauty of the Park by any additional super grid linkage..."

1997
Eryri Local Plan
Deposit Version

- Energy policies
- OHL policy

→

1999
Eryri Local Plan
Adopted Version

- Energy policies
- OHL policy

Energy

...the development of larger scale energy generating schemes... aroused considerable controversy... not only because of their siting within or close to the National Park but also because of the associated high voltage transmission lines...

Overhead Power Lines

Parts of Snowdonia's landscape and in particular prominent views into and out of the National Park have been spoilt by the construction of high voltage transmission lines ... The NPA consider that the visual impact of Supergrid transmission lines is quite incompatible with the aims of conserving the natural beauty of the Park.

...generating companies may wish to construct new Supergrid power lines through the Park to connect a station to the National Grid. The NPA will resist this and insist that new lines are routed around the Park. Where this is not possible, the NPA will insist that Supergrid power lines are laid underground where they need to pass through the Park in a manner which is not detrimental to any environmental interests...

All new connections to the low voltage electricity supply in the National Park should wherever possible be located underground.

The NPA will resist the erection of further overground Supergrid electricity pylons and transmission systems within the National Park.

Where development is proposed adjacent to the National Park, the NPA will request neighbouring Authorities to ensure that the siting does not significantly harm the visual amenity of the Park.

Figure 7.8 Policy development by the Snowdonia National Park Authority

An "energy" chapter outlines the history of energy generation in North Wales and the development of a number of controversial large-scale schemes in and around Snowdonia within the last 40 years. A mention of OHLs constructed to connect the generation projects to the grid is made, with the comment that the OHLs are felt by many to be more intrusive on the park's landscape than the power stations themselves. The following points are made:

- the park requires protection from the effects of large energy installations;
- redundant infrastructure should be removed;
- large-scale energy schemes, including associated transmission systems, are "major development" and so come under a separate policy on major developments (see below).

The energy chapter also contains a specific section on OHLs. This refers again to the existing OHLs within Snowdonia, describing them as detrimental to its landscape and at odds with the purposes of the park. The possibility of applications for further OHLs in the park is then raised, in association with any new power station developments that may occur. These developments are not expected to take place within the park, but are considered a possibility in surrounding areas. This section of the plan therefore states the following:

- the NPA would resist the erection of further OHLs in Snowdonia; lines should instead be routed around the park or placed underground if they must pass through it;
- if any new transmission installations are proposed close to the park, the NPA will request that they do not cause visual harm to the park itself.

Lower-voltage OHLs are also referred to in this section of the plan; in particular, where supply lines are to be upgraded or replaced, they should be made unobtrusive, preferably by being placed underground.

"Transmission systems" are also mentioned in the section in the plan on "major developments"; this resists such developments, laying down strict criteria by which they would be acceptable in Snowdonia, such as having to be "in the British national interest".

Finally, OHLs are referred to in the "landscape" section of the plan as one of the factors that have brought about a significant change in the landscape of the park since its designation.

A number of representations relating to the plan's treatment of transmission installations was made during the early consultation stages of the plan. NG initially objected to the identification of OHLs as a form of major development, quoting a decision by the Secretary of State for the Environment in support of this objection; however, this argument was rejected by the NPA with the assertion that the Secretary of State in question had no jurisdiction in Wales. NG also objected to the NPA's insistence on placing lines in the park underground, referring to environmental, agricultural, technical and financial obstacles to undergrounding. But the NPA resisted this stating, "It is not unreasonable to have a policy protecting the landscape from a form of development which has been universally damaging in the past" (SNPA 1997a). Finally, the Welsh OfficeWelsh Office objected to the

NPA's approach of seeking to prevent new installations outside the park from visually harming the park itself; this too was rejected by the NPA:

> The LPA consider that the siting of Supergrid connections immediately adjacent to the park boundary has caused significant harm to the amenity of the park, in terms of the perception of the park from outside and in terms of their impact on surrounding countryside from within the park (SNPA 1997a).

None of these representations was pursued as far as the plan inquiry. The only issue relevant to HVDT installations discussed at the inquiry was the definition of large-scale energy schemes as a form of major development (SNPA 1997b). The inspector judged this policy to be unnecessary, given that the plan already contained a more general policy on major developments. The other policies referred to in the preceding text proceeded to adoption largely in the form put forward at the deposit stage.

The wording of these policies was not drawn from any preceding plans for the park area. However, similar themes had been expressed previously (Figure 7.8). A non-statutory management plan prepared by the NPA resisted any further energy generation or OHL projects; existing OHLs were described as a "violation of the great natural beauty of the Park" (SNPA 1977). The 1986 review of this plan reiterated this position and proposed steps to ameliorate the visual effects of some of the energy installations, such as the undergrounding or re-routeing of some OHLs (SNPA 1986). Also, conservation policies adopted in 1991 stated the NPA's resistance to new transmission lines (SNPA 1991).

7.12 KEY POLICY ISSUES

7.12.1 History of energy installations in and around Snowdonia

The above policies, and the NPA's defence of them, demonstrate a keen awareness of the historical use of the national park area and immediate surrounds for the location of energy-related installations. This has resulted mainly from the area's mountainous terrain, which has long encouraged the exploitation of water power. Commercial generation of electricity from this source began with the construction of three small-scale hydroelectric schemes in the early decades of the twentieth century. The 1950s saw the expansion of two of these installations and the development of a major pump storage project at Tanygrisiau; a second such project followed at Dinorwig in the 1970s. Two of these schemes lie just outside Snowdonia, while the others are within the park. The largest generation project, however, was the construction during the 1960s of the Trawsfynydd nuclear power station (currently being decommissioned); central Snowdonia proved an attractive location for this development because of its remoteness from main centres of population. Another nuclear power station of the same type was built relatively nearby, at Wylfa on Anglesey.

These schemes inevitably gave rise to the need for power line connections to the grid.

- Major 400 kV lines were constructed in the 1960s to link the two new nuclear power stations to the transmission grid. These lines form a loop, around and through the park. They consist wholly of OHLs, except for a 6 km stretch of underground cable across the Glaslyn estuary (just outside the park).
- A 275 kV OHL was constructed for the Tanygrisiau pump storage scheme; this 7 km line is wholly within the park.
- A 400 kV line connects the Dinorwig scheme to the grid; this 10 km line was wholly undergrounded and lies just beyond the park's boundary.
- A number of 132 kV and lower-voltage OHLs connect the other schemes to the local electricity distribution network. The distribution system also includes other OHLs that straddle sections of the park and its surrounds, to some extent in parallel with the high-voltage lines.

These lines variously cross areas of rugged upland, moorland, hilly farmland, river valleys and woodland; they are, for the most part, highly visible from some of the roads, footpaths and open land in and around the park (Figure 7.9).

Figure 7.9 400 kV OHL crossing the Afon Tryweryn Valley, Snowdonia

7.12.2 Response to energy installations

A considerable number of energy generation projects and associated power lines has therefore been located in the park's area and immediate surroundings over the last century, the most large-scale of these having been proposed and constructed since the designation of the national park. Some of these projects initially found local authority support, because of the economic advantages they offered:

> ...there were obviously economic and social reasons for permitting Trawsfynydd power station back in the early 1960s... rural depopulation, the slate industry was in decline, they needed major civil engineering jobs at that time... culminated with Dinorwig pump storage scheme in the mid 70s... The concern with overhead power lines is they're a necessary evil in many respects (SNPA Planning Officer 2001).

This is a reflection of the weak management of the national parks prior to the establishment of the NPAs, when the parks were in the hands of county authorities not always sympathetic to the aims of designation (MacEwan and MacEwan 1982). Concern was nonetheless expressed about the wider effects of these projects on the park. For example, in 1970, the Snowdonia Joint Advisory Committee drew together an "Electricity Consultative Panel" which sought to negotiate with the electricity utilities' means of ameliorating the "most damaging examples of unsightly routes" of OHLs (Snowdonia Joint Advisory Committee 1971).

However, the creation of the NPAs under the 1972 Local Government Act shifted planning functions to bodies specifically focused on national park purposes. Snowdonia NPA (which took office in 1974) soon made a policy issue of energy installations, expressing its opinion, in the 1977 and 1986 management plans referred to above, that OHLs were detrimental to the park's landscape. The NPA made approaches to the nationalised electricity utility of the time, asking them to improve the appearance of, or even remove, some of their installations, such as pipelines; Snowdonia was perceived to be "the only National Park that has experienced development by the CEGB (Central Electricity Generating Board)" (SNPA 1986).

The establishment of an authority representing national park interests more strongly therefore enabled the initiation of policy resisting large energy installations in Snowdonia. The later requirement upon the NPA to prepare a local plan then provided the opportunity to develop policy further in line with national park purposes.

The NPA found support for its resistance to large-scale energy installations from the reports of committees set up to review the performance of the national parks. Firstly, the Sandford Report noted the anxiety felt about the extent of industrial development, including transmission lines, in the parks, remarking that it was out of accord with park purposes (DoE 1974). Secondly, the Edwards Report noted the extent of "disfiguring power lines" in Snowdonia and the Peak District; this also quoted the view of the Association of National Park Officers that incongruous development, such as high-voltage OHLs, should be removed where possible (Countryside Commission 1991).

There has also been a broader-based, and long-standing, public resistance to energy projects in the park. For instance, the 1950s hydroelectric expansion was originally intended to include a number of other schemes, but these were abandoned following opposition from the public, voiced partly through members of parliament, and from the then National Parks Commission; associated OHLs were specifically mentioned in this regard. Similarly, the 1970s Dinorwig scheme was opposed by the Council for the Protection of Rural Wales and local interest groups; this and county council opposition led to the decision to underground the connection to this scheme (Goulty 1990). Public unease about the presence of major energy installations in the park has continued to the present time; for example, a pipeline to the Cwm Dyli hydroelectric scheme at the base of Snowdon continues to be a focus of dissatisfaction (Figure 7.10) (SNPA Planning Officer 2001). However, public opposition to energy installations in and around the park has not recently crystallised into organised protests or lobbying; it has, nonetheless, formed a background of concern of which the NPA has been aware (SNPA Planning Officer 2001).

Figure 7.10 Pipeline to the Cwm Dyli hydro-electric station at the base of Snowdon

7.12.3 National park status

Central to the NPA's growing concern with major energy installations has been the conviction that they fly in the face of national park designation. Policy has therefore drawn upon provisions seeking to ensure the fulfilment of national park purposes.

This is especially seen in relation to the definition of large-scale energy installations (specifically including high-voltage OHLs) as a form of "major development". This has allowed the NPA to draw on established planning criteria.

Central government and Welsh guidance state that major developments should only be permitted in national parks in exceptional circumstances and sets out the considerations to be taken into account, such as the national need for the development (DoE 1992c, 1997b, Welsh Office 1999). This guidance is based upon the "Silkin test" (suggested during the passage of the National Parks and Access to the Countryside Act 1949 as a guide to mineral extraction in the new parks.) Snowdonia NPA took up the guidance in their policy on major developments (SNPA 1999); an "Energy" issues report made specific reference to the Silkin test (SNPA 1993). The NPA is therefore looking to this provision as a means of protecting the park from unwelcome energy; the importance to the NPA of being able to categorise OHLs as a form of major development is indicated by their rebuttal of NG 's objection on this point.

The perceived incompatibility of major energy installations with national park designation is also illustrated by the effects that OHLs are felt to have on the landscape of Snowdonia. The park's landscape is understood to be of outstanding quality, described in the plan, for example, as "spectacular" and "dramatic", consisting of "unspoilt… scenery and views" (SNPA 1999, p. 10, 14). The openness and relative wildness of the mountain and moorland areas are thought of as vulnerable to the visual effects of OHLs, which are seen as having a contrasting artificial nature:

> Some of the pylon routes are very prominent on skylines; they are significant large man-made objects ... People come here to walk mountains; we have extensive moorland areas, sweeping views and very little vertical elements to them … I don't think people want to walk up hills to be assaulted by a great row of pylons marching off into the distance, or maybe wind farms on the opposite hill. I think people need a sense of solitude and wilderness with no man made objects in a landscape - this type of area is more and more difficult to find today … (SNPA Planning Officer 2001).

This officer's opinion indicates priority being given to the retention of a landscape free from OHLs (and wind turbines) for the benefit of walkers seeking an increasingly rare experience of "wilderness". It implies that OHLs run counter to two of the statutory purposes of Snowdonia: the conservation of the natural beauty of the park and the enjoyment of its special qualities by the public (Environment Act 1995):

> … national parks are the highest landscape designation we have in this country, there must be purposes for that … there should be some areas of the country which do remain a bit wild, and you avoid this man-made intrusion into this landscape … we're a national park for a purpose, and I think one of the purposes is to try to preserve little pieces of wilderness, or semi-wilderness, where we can (SNPA Planning Officer 2001).

The importance to the NPA of preserving the park's characteristics is also illustrated by their insistence that development immediately *outside* their area should also respect the amenity of the park.

7.12.4 Policy precedents

The plan presents two means of avoiding further visual intrusion from OHLs: any new lines should be routed so that they avoid the park altogether or, if this is not possible, placed underground. Both of these techniques have, in the past, been carried out in respect of the park: the 400 kV line was deliberately routed outside Snowdonia's northwest border and sections of high-voltage lines were undergrounded across the Glaslyn estuary and where they connected with the Dinorwig scheme. These steps were undertaken specifically with the park's status in mind; the Glaslyn estuary undergrounding, for instance, was carried out in order to preserve a classic view towards Snowdon (Goulty 1990). These cases have therefore acted as precedents of measures capable of ameliorating the effects of OHLs; the Glaslyn example, in particular, has provided the NPA with a model that has influenced policy (SNPA Planning Officer 2001).

7.12.5 Possible future energy developments

The policies resisting major energy installations are set within the context of possible future applications for new generating schemes and associated OHLs in or around the park. Such applications are seen as a real possibility, especially for projects exploiting renewable sources of energy. During the early stages of plan preparation, the NPA gave consideration to "likely pressures" of this kind in an "Energy" issues report (SNPA 1993); this followed on from the Gwynedd Renewable Energy Study (Chris Blandford Associates 1994), which identified wind, water and biomass as the renewable sources with the greatest potential in North Wales. The plan picks up on this theme, also mentioning the encouragement given to renewable energy schemes by the Non-fossil Fuel Obligation; this is described as having "created a somewhat uncoordinated rush to capitalise on the renewable premium and resulted in a plethora of applications for renewable energy developments" (SNPA 1999).

The plan's concern for the environmental effects of new energy projects is therefore a response, in part, to increasing pressures for energy to be derived from renewable sources and to the awareness that the region contains considerable potential in this regard. This is illustrated by the NPA's opposition to wind farms being located in or close to the park, for similar reasons as for OHLs:

> The National Park Authority is very strongly against wind farm developments ... You're getting tower heights in excess of 300 feet now ... the impact of those on our hillsides would be completely detrimental ... There have been no applications, as far as I know, for windfarms in any national park ... There have been round the

> fringes and we obviously have made representations and objections to a number of those ... (SNPA Planning Officer 2001).

As well as seeking to defend the park from the effects of new energy schemes, the NPA also seeks the removal of redundant "energy plant, structure and transmission systems once they are no longer operational" (SNPA 1999); this aim relates partly to the decommissioning of Trawsfynydd nuclear power station, which has led to hopes that it would be possible to dismantle it and some of the associated OHLs (SNPA Planning Officer 2001).

7.13 CONCLUSION

Throughout its 50-year history, Snowdonia National Park has been affected by a number of energy generation schemes that have exploited the natural terrain and the area's relative remoteness; some of these schemes have had significant effects on the very landscape that gave rise, in part, to the park's original designation. The construction of connecting lines has been one of the most serious of these effects, and has provoked considerable controversy and opposition to energy projects. Various national park bodies have been involved in discussion about these and other proposed projects and have echoed wider concerns about the effects of the various schemes on Snowdonia. Energy installations have therefore been a long-standing issue of immediate relevance to land-use planning in the park. This has been reflected in policy formation by the NPA since its inception; the establishment of a body specifically focused on national park purposes has encouraged the development of policies resisting major energy installations.

A number of other factors have also contributed to the NPA's formation of policy:

- government guidance on major developments in national parks has provided the NPA with supportive planning criteria;
- national park review committees have added weight to the NPA's contention that OHLs are damaging to the park's landscape;
- the landscape character of the park is such that OHLs are perceived as detracting from its qualities;
- measures taken locally to ameliorate the effects of OHLs have provided precedents that have gained policy expression;
- the shift towards renewable energy generation and knowledge of the area's resources in this regard have raised the possibility of proposals for further energy schemes in the area.

The components of policy formation by Snowdonia NPA are depicted in Table 7.3. The inter-relation of the principal components and the progression of policy through the DPP are summarised in Figure 7.11.

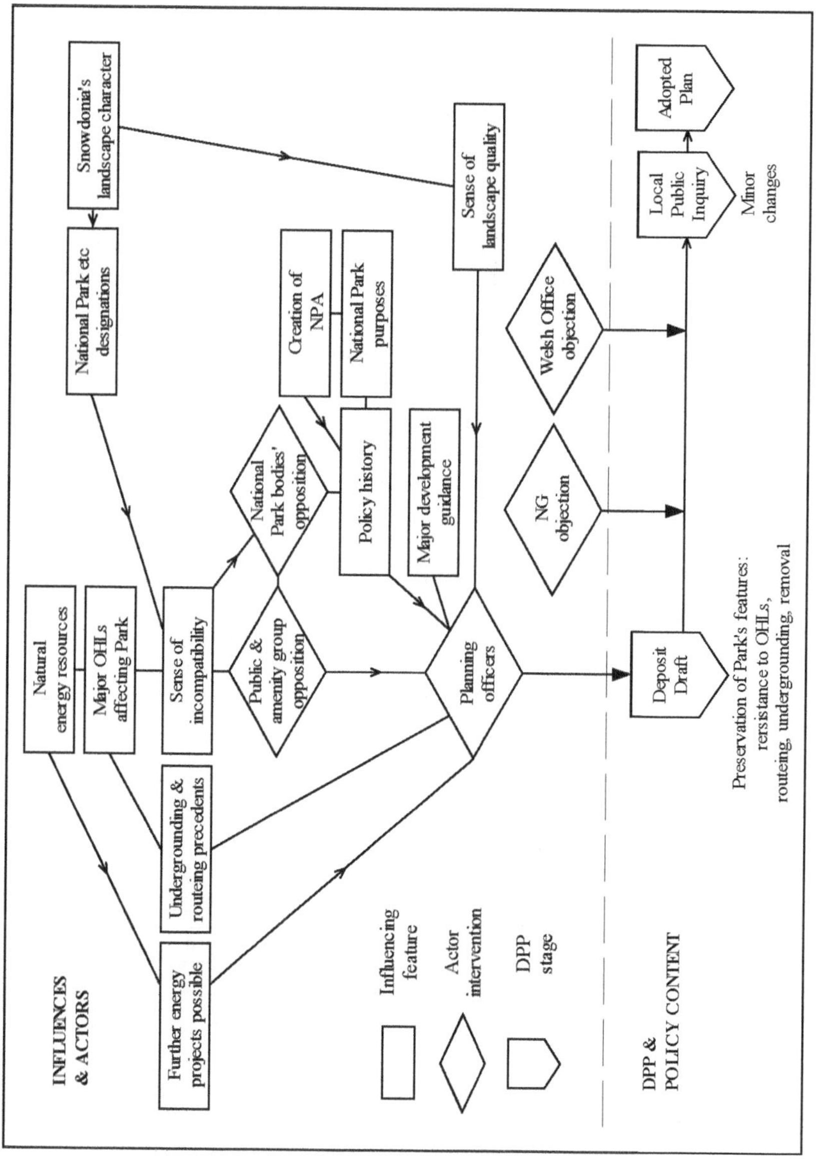

Figure 7.11 The process of policy formation by Snowdonia National Park Authority

Table 7.3 Components of policy formation by Snowdonia National Park Authority

BROAD CATEGORIES		WITHIN-CASE COMPONENTS	CROSS-CASE COMPONENTS*
INFLUENCING FEATURES	GEOGRAPHICAL, ECONOMIC AND SOCIAL FEATURES	Regional renewable energy pressures	Pressure for energy projects
		National park and many other designations	Extensive designated areas and sites
		Extensive rural areas	Extensive rural areas
		Renewable and other energy resources	Potential for energy generation projects
	HVDT INSTALLATIONS (AND TELECOMMUNICATIONS)	Major OHLs within and around the park	Prominent installations in countryside
			Installations affecting designated areas/sites
		Likelihood of future renewable energy-related powerlines	Perceived likelihood of future HVDT projects
	POLICY FRAMEWORK	National park purposes; guidance on major developments	Supportive national planning policy
		Opinion of national park bodies and review committees	National committee/association support
	PLANNING EXPERIENCE	1970s and 1980s park plans	History of HVDT-related policy
		Routeing and undergrounding precedents	Local precedents of measures to offset effects of HVDT
	PLAN THEMES	Sustainability	Sustainability
		Conservation of park's features	Protection of environmental features
		National park purposes	National park purposes
	INSTITUTIONS	Creation of NPA	New institutional arrangements
ACTORS	PLANNING OFFICERS	Emphasis on national park purposes	Strongly supportive of policy
		Sense of park's landscape character and quality	Landscape valuation/assessment
	INTEREST GROUPS	Long-standing public and amenity group opposition to energy schemes and associated OHLs affecting the park	Opposition to HVDT projects
			Landscape, etc. protection concerns
	PUBLIC		Opposition to HVDT projects
			Landscape, etc. protection concerns
	ELECTRICITY INDUSTRY	Opposing proposed policy	Opposing aspects of policy
DPP	REPRESENTATIONS	NG and Welsh Office opposing aspects of proposed policy	Opposing HVDT-related policy
	LPA DEFENCE	Rejection of objections	Strongly defending policy against objections
	Local public inquiry (LPI)/Examination in public (EiP)	Inspector recommended alterations	Inspector/Panel recommended alterations

BROAD CATEGORIES		WITHIN-CASE COMPONENTS	CROSS-CASE COMPONENTS*
POLICY CONTENT	BROAD THEMES	Incompatibility of OHLs and park	Countryside protection
	EFFECTS OF HVDT INSTALLATIONS ON:	Park's landscape, conservation value and amenity	Visual quality: countryside
			Designated areas: national
	MEASURES TO OFF-SET EFFECTS	OHLs resisted as major devel-opment	Opposition to new HVDT installa-tions
		Removal of redundant infrastruc-ture	Removal/re-routeing of existing OHLs
		Routeing lines around park	Careful routeing of new OHLs
		Undergrounding through park	Undergrounding of lines

* These refer to components defined for the case studies as a whole; see Section 9.2.1.

CHAPTER 8

Case Studies (3): Town – Rural Areas

8.1 INTRODUCTION

The case studies in this chapter explore the concerns of local planning authorities (LPAs) in areas containing towns in rural settings. Redcar & Cleveland Borough Council's area divides between industrialised towns and attractive countryside. Mendip District Council covers a rural area containing a number of small towns. Aylesbury Vale District Council and Bridgend County Borough Council both cover mostly rural areas with dominant towns at their centres. These four authorities all raise a range of HVDT-related concerns; they deal both with questions of the proximity of HVDT installations to other development and with the protection of designated areas and sites (or of the wider environment) from HVDT installations. These case studies are characterised by a wide range of features contributing to policy formation, within which local planning cases play a significant part.

REDCAR & CLEVELAND BOROUGH COUNCIL

8.2 INTRODUCTION

Redcar & Cleveland Borough Council (RCBC) is a unitary authority in Cleveland in northeast England. The area includes major sites of heavy industry on Teesside, where the decline of the traditional petrochemical and steel industries has led to problems of unemployment, depopulation and dereliction. In marked contrast, the borough also covers large areas of attractive countryside rising up from the coast to the North York Moors National Park, much of which has received local or national designation for landscape or conservation interest.

RCBC was established as a unitary authority following local government reorganisation in 1996, being the successor to Langbaurgh on Tees Borough Council (LTBC). LTBC was a district council within the now abolished Cleveland County Council. Although a unitary authority, RCBC prepares a local plan (LP), under the 1995 Cleveland Structure Plan. A review of the structure plan (SP) is being carried out by the Tees Valley Joint Strategy Unit on behalf of five unitary authorities.

The council adopted its LP in June 1999. It seeks to encourage regeneration in order to tackle unemployment and deprivation and to sustain the borough's population level. Set alongside these objectives are concerns for environmental protection, which aim partly to contribute towards improving the image of the borough. This case study was carried out during the period following adoption.

8.3 ARTICULATION OF POLICY

A policy relating to HVDT installations was brought through plan preparation, as begun by LTBC and completed by RCBC, under "Services, infrastructure and constraints". It is entitled "overhead lines" and is primarily concerned with the protection of designated and other areas from the visual intrusion of overhead lines (OHLs) and other installations (Figure 8.1).

This OHL policy finds its earliest expression in planning policy guidance (PPG) produced by LTBC in 1991. This stated that the council would normally oppose the siting of OHLs in certain designated areas, where residential areas would be adversely affected, and across recreation areas (LTBC 1991). The main justification for the guidance was to protect the visual environment of the borough, though in the case of residential and recreation areas, health and safety issues were also raised.

This guidance provided a precedent when preparation of the LP began in 1993. The deposit draft (LTBC 1995) contained a policy, which followed much of the wording of the council's 1991 guidance and made specific reference to it. The policy carried forward the same concern for the protection of areas and sites designated for landscape, conservation, historic and archaeological reasons; areas recognised both nationally and locally were included without distinction. With regard to residential areas, health concerns were again raised, though with an acknowledgement of the advice of the National Radiological Protection Board (NRPB). With regard to recreational areas, safety issues were again cited as a reason for opposing OHLs, though National Grid's (NG's) safety measures were recognised. These modifications to the planning guidance were made in response to representations by the electricity industry during the plan's early consultation period.

Further representations were made against the policy by the electricity industry, both NG and Northern Electric, along the following lines:

- the extent of areas from which installations would be excluded was so great as to be an almost "blanket opposition" to OHLs in the borough; in particular, exclusion from locally designated special landscape areas would prevent them from providing an efficient electricity network.
- the industry's own practice of minimising the environmental effects of their activities are adequate measures to mitigate the visual effects of installations.

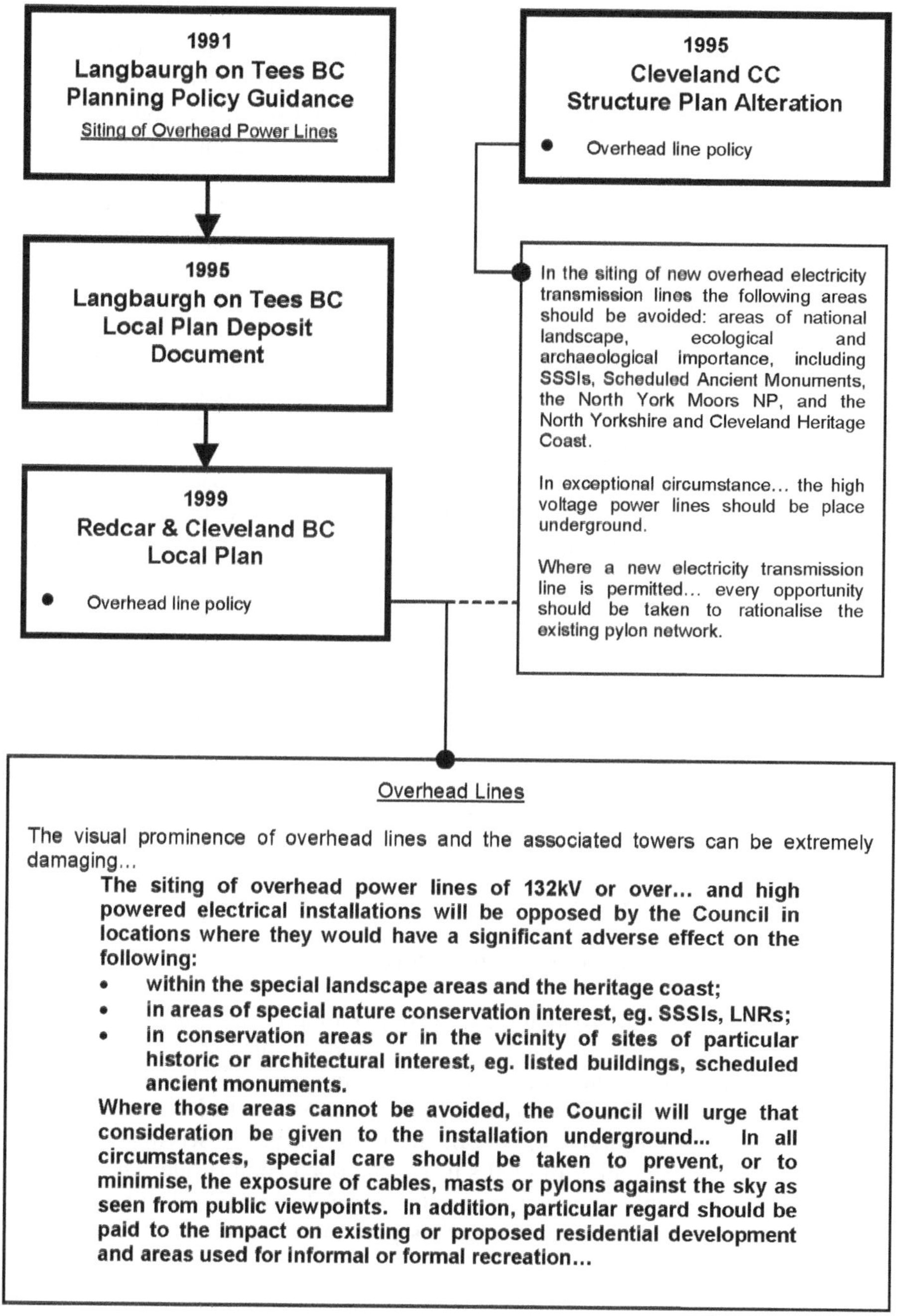

Figure 8.1 Policy development by Redcar & Cleveland Borough Council

These arguments were carried forward to the plan inquiry in 1997.

At the inquiry, the council (now RCBC) defended its position as follows:

- the industry's duties are limited in the degree of environmental protection they can provide. For example, the 'Holford Rules' on careful line routeing (Section 2.10.2) are non-statutory and not subject to public consultation – unlike the council's plan;
- Regional Planning Guidance for the Northern Region (DoE 1993a) and the Cleveland Structure Plan (Cleveland County Council 1995) both include policies supporting the protection of designated areas from visual and other impacts of OHLs; the SP policy especially parallels that of the council in some respects (see Figure 8.1).

During their evidence at the plan inquiry, both sides made reference to a public inquiry held in 1992 to consider NG's proposal for a major OHL to be constructed from Teesside to York (the "North Yorkshire" line). NG made the point that the inspectors had recommended routeing part of this line through a special landscape area within Redcar & Cleveland, despite the designation. The council stated that the inspectors had also recommended the undergrounding of a section of the line through a locally designated landscape area, thus recognising that the designation did warrant protection for the area from the visual intrusion of OHLs.

The plan inquiry inspector supported the inclusion in the plan of an OHL policy, but recommended that the unqualified opposition to OHLs in all the specified areas should be modified to a less restrictive approach, whereby OHLs would be opposed "where they would have a significant adverse effect". The inspector also recommended that where such locations could not be avoided, consideration should be given to placing lines underground. He did not suggest, however, that any distinction should be drawn in the level of protection sought for areas designated nationally or locally. On the matter of health concern, he recommended its deletion from the policy. The council accepted these recommendations.

8.4 KEY POLICY ISSUES

8.4.1 Broad concern regarding overhead lines

The council's initiative in producing their own planning policy guidance on OHLs in 1991 is an indication of a relatively long-standing interest in the effects of OHLs. Their concerns should be seen in the context of the density of HVDT installations in the borough. These include five NG-owned substations and associated OHLs in the urban section of the borough, partly serving Teesside's heavy industry, and a number of OHLs traversing open countryside. These installations are highly visible in certain areas (Figures 8.2 and 8.3). The presence of this equipment has raised concern about its possible effects, particularly among elected members:

> ... there is a view that ... we do have the visual impact, we do have the potential health problems. I do recall that they [members] were very concerned about the unknown health effects around schools,

around residential areas, around play areas... so they looked at it from both angles really (RCBC Planning Officer B 2001).

Figure 8.2 HVDT installations near housing, Lackenby, Middlesbrough

Figure 8.3 OHLs on the Eston Hills, Cleveland

These different aspects are looked at in more detail in the following paragraphs. Firstly, however, the effect of a controversial project on overall attitudes to OHLs in the borough is considered.

The "North Yorkshire" line

This line is currently being constructed by NG. It is a 400 kV line 75 km long and will be the largest addition to the transmission system since privatisation of the electricity industry. It is being developed to connect a new power station on Teesside, in RCBC's area, to the grid. The application to construct the line was made in 1991, and it quickly generated local controversy (Sheate 1995). It was opposed by the LPAs affected, including RCBC, and was subsequently made the subject of two lengthy public inquiries in 1992 and 1995. Consent for the final part of the project was granted in 1998; this included a requirement to underground a section of the line, as referred to above. RCBC's plan preparation fell largely within this period.

Public opposition to the line was based on a number of objections, not least the impact on the rural character of the area that it would pass through. Opposition became organised through the formation of a determined interest group, REVOLT (Rural England Versus Overhead Line Transmission) , which continues to protest against the line during the course of its construction. This opposition has not been at its most intense within RCBC's area, but the level of controversy has heightened public awareness of OHL issues in the borough:

> I know that the line through Hambleton has generated a lot of press; if you drive up the A19 … there are farmers that have put in fields great big fertilizer containers painted yellow with pictures of pylons on, saying no more pylons here, so it's quite an emotive issue locally (RCBC Planning Officer A 2000).

Moreover, the "North Yorkshire" line has had specific implications for the council, as the section of the route which does pass through their area, though only about 5 km long, crosses a special landscape area (illustrated in Figure 8.3).

In addition, other local authorities affected by the line have been bringing forward comparable policies (Figures 5.7–5.10). These policies have not been advanced in a coordinated manner, although some informal consultation has taken place between authorities; for example, RCBC tracked the progress of Hambleton District Council's policy, as it was in advance of theirs by about a year.

The power station that has created the need for the "North Yorkshire" line is a combined-cycle gas turbine plant and has been built on Teesside to take advantage of the North Sea gas pipeline terminal. It was one of the United Kingdom's earliest "independent power producers", producing energy in order to sell it to the electricity "pool" and results directly from liberalisation of the electricity industry (Patterson 1999). When consent was granted for the power station, however, the subsequent need for a new link to the grid was not fully appreciated:

> The local authorities got caught out … We were consulted on the application for the power station, but the implications of what that

> meant were not spelt out … We sort of assumed it would just feed into the existing net ... but then the application came in for … new transmission lines, and they said, oh well, but you didn't object to the power station (former Cleveland County Council Planning Officer 2001).

This experience, along with the possibility of further applications for power stations, provoked the then Cleveland County Council to introduce policies to its SP, guiding its response to possible future applications for power stations and OHLs (including the policy shown in Figure 8.1). (Reaction to the "North Yorkshire" line experience can also be seen in the SP replacement, which calls for applications for power stations and related OHLs to be considered at the same time (Tees Valley Joint Strategy Unit 2000).)

Part of the opposition to the line was a feeling that it was designed for the export of power to the south of England, and that a local disbenefit was being created (Lomas *et al.* 1996). With regard to Teesside, this view correlates to longer-standing attitudes towards the wider presence of heavy industry:

> … there's certainly an attitude that we get a lot of these nasties dumped upon us, and that certainly has been the thinking with the overhead lines, because the power stations are here because of oil coming in from the North Sea, then the power has to be transported elsewhere, so … very often we don't get to use it ourselves (RCBC Planning Officer B 2001).

Even though RCBC's policy was taking shape before the "North Yorkshire" line was applied for, the controversy over this line has influenced the council's stance in a number of ways:

- it heightened the council's pre-existing concerns about the effects of OHLs, creating a public debate on the proposed development and the possibility of further proposals;
- the application to route part of the line through the borough increased the relevance of OHLs;
- the line's route through one of the special landscape areas intensified the borough's aim of protecting designated areas;
- the line added to a sense of suffering the consequences of infrastructure that bring little immediate benefit to the area;
- as other authorities also took up the concerns being expressed, they provided RCBC with a measure of support for their policy.

8.4.2 Protection of landscape and conservation interests

The main focus of RCBC's OHL policy is the protection of designated areas; these extend across a significant proportion of the borough and include:

- special landscape areas, covering much of the inland area;
- heritage coast, along the Cleveland coastline;
- scheduled monuments, especially a concentration of archaeological sites;

- conservation areas, in historic town and village centres;
- wildlife conservation sites.

These designations indicate the value that is attached to landscape and other local features. There is a perception that the borough contains areas of exceptional quality, particularly within the context of the Tees Valley:

> ... within the Tees Valley, we've got the highest quality of countryside of the five boroughs, and that's recognised by the special landscape status; we've also got more archaeological remains, more conservation areas, so although we've got some very poor areas from the point of view of the environment, we've also got some of the very best (RCBC Planning Officer B 2001).

This view, evident among elected members and officers, has motivated a strong stance on protecting designated areas from development; the plan contains a considerable number of policies aiming at protecting them. The importance attached to landscape conservation is also shown in an appendix to the plan, which describes landscape character types in the area with the intention that development should respect these different identities (RCBC 1999). The OHL policy is consistent with the council's strategy towards these areas:

> [Members] accepted that we couldn't refuse everywhere, because obviously we have large electricity producers in the borough, and they need to be able to get that onto the grid somehow, but ... they wanted us to try and avoid if possible any detrimental effect to some of our landscape (RCBC Planning Officer B 2001).

The sense of need to protect the designated areas from OHLs also arises from the perceived pressure for future lines. The designated areas are mostly located between the highly industrialised zone of Teesside and the North York Moors National Park. Teesside is a source of possible OHL development, as evidenced by the case of the "North Yorkshire" line. The National Park, ironically, also produces pressure upon areas surrounding it, because developments that would not take place within its borders may be placed just outside. "We're very concerned that we don't get the overspill from the park" (RCBC Planning Officer B 2001). There is a sense of squeeze upon the borough's designated areas, leading the council to seek to protect them from developments such as OHLs.

An additional concern for the visual appearance of the borough arises from the council's regeneration efforts. For example, the plan speaks of the extent of derelict land on former industrial and mining sites, and of the need to deal with "derelict land and the associated poor image of parts of the borough" (RCBC 1999, p. 4). Although not directly related, the OHL policy can be seen as contributing to enhancement of the environment in the interests of economic regeneration. "We get anecdotal reports from small businesses ... that the presence of overhead lines does impact on their business, in particular those out in the rural area" (RCBC Planning Officer B 2001).

Finally, the existence of a supportive policy framework encouraged RCBC to pursue this aspect of its policy; Regional Planning Guidance and the county SP provided a background of policy similar to the council's approach. The 1992 public inquiry into the "North Yorkshire" line, although not entirely favourable to the council, also provided some backing in requiring undergrounding of the line, in part, of a locally designated area.

8.4.3 Protection of health and safety

The earlier versions of OHL policy sought to exclude new lines from residential and recreational areas on the grounds of health and safety fears:

> Such lines are opposed close to residential areas, again because of their visual intrusion and concern over the effect on health and welfare of residents living close to the lines. Wherever possible, despite the safety measures always taken by National Grid, it is felt that location close to recreation areas should be avoided (LTBC 1995).

Pressure for these references to be included in the policy came from elected members, despite the advice of officers that there was little justification for this approach. Members were responding, to some extent, to public anxieties about the possible effects of electricity installations in their vicinity:

> About health concerns: those are things that there is no evidence for at the moment ... but we still can't stop people, the general public, from having fears and concerns ... certainly that was an issue that has been raised at committee ... It's been spelled out to members that we can't have a policy relating to health ... (RCBC Planning Officer A 2000).

However, member interest in this issue also stemmed from their wider interest in health matters, which is related to the industrialised character of the borough:

> Our members ... have a very real concern about health risks ... and actually long before they ever got involved in electricity pylons, they had a very good track record on things like asbestos ... it is borne out of very genuine concern, and of course we are in a geographical location where we are subject to quite a number of proposals ... (RCBC Planning Officer B 2001).

The council did not succeed in retaining any references to health and safety in the adopted policy; the policy does still oppose the siting of OHLs near residential and recreational areas, but only on the grounds of visual intrusion and unspecified "impact" (RCBC 1999). However, the degree to which the council had attempted to defend their position on this and other aspects of the policy is indicated by the fact that they chose to be professionally represented for this part of the plan inquiry.

8.5 CONCLUSION

RCBC's process of OHL policy formation has been orientated towards protection of the well-being of residents and the quality of rural areas. These twin interests reflect the contrasting nature of the borough, from urban, industrialised Teesside to the swathe of attractive countryside.

These concerns resulted initially from the existing high density of HVDT installations in parts of the borough and the perception of possible future developments. This has provoked

- public and member anxieties about the possible health and safety implications of OHLs, particularly in the urban areas; this has been related to members wider interest in health matters;
- a resolve to protect the large areas of designated land from future OHLs; this has been encouraged by the local value attributed to these areas and by the sense of pressure upon them due to the proximity of Teesside and the National Park.

These concerns were given extra impetus by the controversy over the "North Yorkshire" line. This project

- raised the profile of OHLs and their perceived effects in the locality;
- brought OHL applications more directly into the council's planning arena;
- intruded upon one of the designated areas;
- generated policy formation within other authorities, thus producing a policy framework that was supportive to the council. In addition, the siting of the power station that led to the need for the line is seen as a possible precursor for other similar developments; the liberalisation of the electricity industry therefore provides a wider background to policy formation.

Finally, the policy is in line with a strategy of enhancing the environment in the interests of economic regeneration and countering the sense of environmental disbenefit resulting from major industry. The components of policy formation by RCBC are depicted in Table 8.1. The inter-relation of the principal components and the progression of policy through the development plan process (DPP) are summarised in Figure 8.4.

Table 8.1 Components of policy formation by Redcar & Cleveland Borough Council

BROAD CATEGORIES		WITHIN-CASE COMPONENTS	CROSS-CASE COMPONENTS*
INFLUENCING FEATURES	GEOGRAPHICAL, ECONOMIC AND SOCIAL FEATURES	Liberalisation of electricity industry: new power station	Privatisation/liberalisation of electricity industry
		Designated areas and sites in countryside	Extensive designated areas and sites
		Extensive rural areas	Extensive rural areas
		New York Moors National Park adjacent	Proximity of National Park
		Industrial decline and dereliction on Teesside	Economic decline/poor environmental conditions
		Hazardous industries on Teesside; sense of environmental disbenefit	Concentration of hazardous industries
		North Sea gas pipeline terminal	Potential for energy generation projects
	HVDT INSTALLATIONS (AND TELECOMMUNICATIONS)	High density of installations in urban area and traversing countryside	High density of installations
			Prominent installations in countryside
		OHL through special landscape area	Installations affecting designated areas/sites
		North Yorkshire line	Recent/current HVDT projects
		Possible installations for future power stations	Perceived likelihood of future HVDT projects
	POLICY FRAMEWORK	Regional Planning Guidance (RPG) for the northern region, Cleveland structure plan	Supportive regional/structure planning policy
		PPG policy on OHLs	Supportive local planning instruments
		North Yorkshire line inquiry recommendation	Supportive HVDT-related public inquiry decisions
		Nearby LPAs advancing comparable local plan policies	Comparable neighbouring authority policy
		Recognition of NRPB advice	NRPB advice
	PLANNING EXPERIENCE	PPG policy on OHLs	History of HVDT-related policy
		North Yorkshire line	Recent/current HVDT projects
		North Yorkshire line inquiries	HVDT-related public inquiries
		North Yorkshire line undergrounded section	Local precedents of measures to offset HVDT effects
		Tracking nearby authorities policy formation	Knowledge of other LPAs experience
	PLAN THEMES	Environmental protection	Protection of environmental features
		Encouraging regeneration and enhancing the environment	Regeneration/environmental improvement
ACTORS	PLANNING OFFICERS	Actively pursuing policy	Strongly supportive of policy
		Landscape considered best in Tees valley	Landscape valuation/assessment
		Advising against health references	Caution against inclusion of EMF health concerns

BROAD CATEGORIES		WITHIN-CASE COMPONENTS	CROSS-CASE COMPONENTS*
		Contact with nearby authorities pursuing comparable policies	Communication with officers from other authorities
	ELECTED MEMBERS	Strongly supportive of policy	Strongly supportive of policy
		Protective of designated areas	Landscape, etc. protection concerns
		Members wider health concerns and responding to public anxieties	EMF health/amenity concerns
	INTEREST GROUPS	REVOLT: opposition to North Yorkshire line	Opposition to HVDT projects
			Landscape, etc. protection concerns
	PUBLIC	Opposition to North Yorkshire line and health concerns	Opposition to HVDT projects
			Landscape, etc. protection concerns
			EMF health/amenity concerns
DPP	ELECTRICITY INDUSTRY	Objections to proposed policy	Opposing aspects of policy
	REPRESENTATIONS	Electricity industry objecting to policy	Opposing HVDT-related policy
	LPA DEFENCE	LPA defending policy at LPI	Strongly defending policy against representations
	Local public inquiry (LPI)/Examination in public (EiP)	Contestation	Contestation
		Recommended alterations	Inspector/Panel recommended alterations
POLICY CONTENT	BROAD THEMES	Protection of designated areas	Countryside protection
		Protection of residential areas	Residential area protection
	EFFECTS OF HVDT INSTALLATIONS ON:	Visual quality: designated areas	Visual quality: countryside
		Visual quality: residential areas	Visual quality: urban areas
		Nationally and locally designated areas	Designated areas: national
			Designated areas: local
		Conservation sites	Ecological interests
		Archaeological sites	Archaeological sites
		Health: EMF	Public health (EMF)
		Safety: areas	Public safety
		Residential and recreational areas	Residential, etc. areas
			Recreational areas
	MEASURES TO OFFSET EFFECTS	Opposition to new installations	Opposition to new HVDT installations
		Avoiding installations on skylines	Careful routeing of new OHLs
		Undergrounding in designated areas	Undergrounding of lines

*These refer to components defined for the case studies as a whole; see Section 9.2.1.

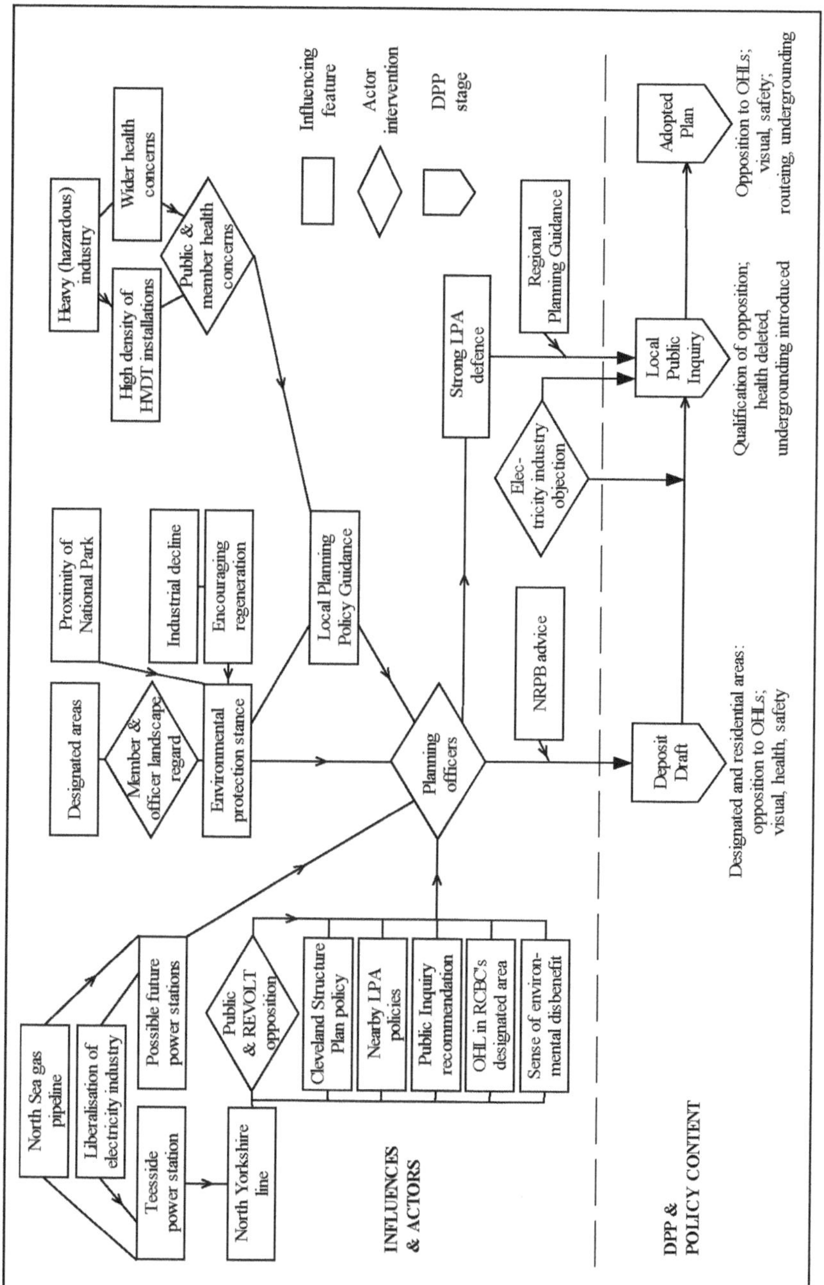

Figure 8.4 The process of policy formation by Redcar & Cleveland Borough Council

MENDIP DISTRICT COUNCIL

8.6 INTRODUCTION

Mendip District Council (MDC) is a district authority in Somerset in southwest England. The area is mostly rural, with a large number of villages and some small towns. The largest settlement is Frome; other prominent towns include Wells and Glastonbury, which have tourist importance. The district includes areas of distinctive character, such as the Somerset Levels, and contains many areas and features designated for their landscape, wildlife, archaeological or historic interest. The authority is currently preparing its first district-wide LP; this case study was undertaken during the period following the public local inquiry into the plan.

The plan sets itself a strategy of sustainable development, in which economic growth and a better quality of life are sought alongside environmental protection. Central to this approach is the concept of environmental capital; the natural and built environment is seen as a valuable resource that is to be both conserved and enhanced.

The plan also sets out provisions for new housing in response to the Somerset structure plan allocation for the district. The council had successfully objected to the draft SP's housing proposals for the district, and the LP now seeks to restrict most new development to the towns.

8.7 ARTICULATION OF POLICY

Policies which the MDC see as relevant to HVDT installations were introduced into the deposit draft version of the LP in the section "Building in Quality", which deals with standards of design for new developments:

1. A policy that seeks to protect designated areas and recreational and residential sites from the adverse impacts of OHLs and other electricity installations. Undergrounding is called for where a power line has to cross a sensitive area.
2. Two other policies in the same section of the plan, on hazardous installations, do not specifically refer to electricity installations, but are also intended by the council to be of possible relevance to OHLs. These policies prohibit the siting of hazardous installations where this would pose a risk to health and safety, and the siting of new development near existing hazardous installations.

The main elements of these three policies are set out in Figure 8.5 (MDC 1998).

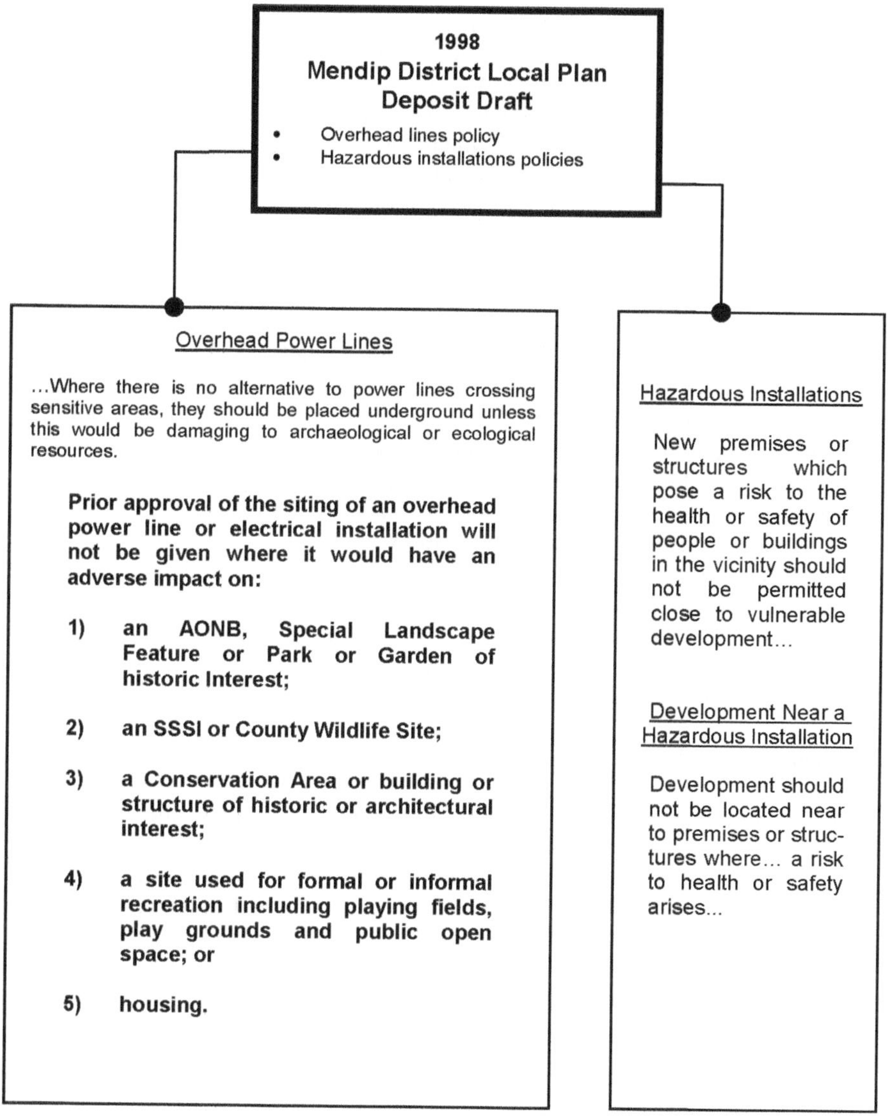

Figure 8.5 Policy expression by Mendip District Council

The OHL policy lists those areas for which protection is sought; areas that are designated nationally or locally and recreational sites and housing, are all to be protected from "adverse impact". NG made representations against this policy, calling for a more scaled approach such that the council would not be seeking the same level of protection for residential and recreational sites as for designated areas. The council accepted this, and proposed changes to policy such that adverse impacts of OHLs need only be "minimised" within residential and recreational areas, and only impacts on "amenity" are specified. NG also objected to the policy's requirement for lines to be placed underground in sensitive areas, pointing out the technical, financial and environmental difficulties involved. This was also accepted by the council, which amended the policy so that undergrounding would only be sought "exceptionally".

Other representations on the OHL policy were made by the Countryside Agency, which sought to extend the policy so that it also protected "countryside and settlement character", and by the British Wind Energy Association, which sought to narrow the policy by having it refer only to "unacceptable" adverse impacts. Both these approaches were rejected by the council.

The hazardous installations policies seek to prevent health and safety risks arising from the proximity of certain installations to other developments. These policies do not mention HVDT installations; instead, one of the policies gives as examples of the kind of installation in mind "explosive stores, fertilizer stores and industrial premises where hazardous substances are used" (MDC 1998). The council considers these policies to be potentially applicable to the perceived health effects of electromagnetic fields (EMF) (see the following text). No representations were made against these policies by the electricity industry, as they would not concede that their installations pose hazards of this kind.

8.8 KEY POLICY ISSUES

8.8.1 Protection of landscape and related features

The OHL policy is concerned firstly with the protection of areas and sites recognised for their landscape, wildlife, historic or architectural importance. The plan places stress on the presence and conservation of these features in the district. For instance, in relation to landscape, a significant proportion of the district is covered by national or local designations, especially Areas of Outstanding National Beauty (AONB)s, Special Landscape Features and Listed Historic Parks and Gardens. A number of policies aim to conserve these, stating, for example, "Development will not be permitted where it would have an adverse impact on the scenic value or integrity of the Special Landscape Features" (MDC 1998). Moreover, the council has commissioned a comprehensive landscape assessment, which defines character areas across the whole district; an associated policy expects development to respect the landscape characters indicated.

These features are described as "environmental assets", to be conserved and enhanced in the interests of the quality of life for residents, and also for wider economic reasons. They are seen as essential to the attraction of visitors and businesses:

> The natural environment is especially important for a rural area like Mendip District. It provides the underlying framework for a healthy

> and viable rural economy. The Plan will seek to protect the environment, not only for its own sake, but because it constitutes an important local asset … (MDC 1998).

Although this strategy is primarily one of protecting environmental assets from the adverse effects of built development, it also provides a context for the OHL policy. Existing OHLs already impinge upon areas of landscape and conservation importance in the district. In particular, a 400 kV line runs the entire length of the district through the Somerset Levels, undulating countryside and the Frome valley. The line is visually prominent along much of its length, particularly upon the low-lying levels (Figure 8.6), on hill rises and as it crosses the Frome valley bottom.

Moreover, disquiet about the effect of OHLs on the local landscape has been expressed within the district. Although this has not been a widespread concern of the general public, it has been raised by groups and individuals with an interest in the countryside, especially the Council for the Protection of Rural England (CPRE):

> … the first people probably to mention it to us was CPRE in relation to the big pylons in the open countryside … The average man in the street on an average day is not aware of them; the countryside lobby and pressure groups are aware of them, but … I think they're regarded as a necessary feature of the landscape, even if they're ugly, by most people most of the time … There is an element that finds them a problem. … the Mendip hills is a potential area for wind turbines which would then of course lead to power lines, I think they see that as a threat (MDC Planning Officer A 2001).

Although CPRE did not lobby for a policy restricting the siting of OHLs (CPRE Mendip Group Former Member 2001), council officers have taken into account their views and those of a wider environmental grouping (see the following text) when preparing the policy. The OHL policy has therefore been formulated against a background of public unease with the effects of OHLs upon the landscape and within a broader policy context of protecting environmental assets. The policy has not been advanced in anticipation of any applications for major OHLs, but with the possibility of new lower-voltage lines in mind, conceivably related to renewable energy projects (MDC Planning Officer A 2001).

Preparation of the current plan provided the council with the opportunity to introduce a policy on OHLs. The council did have previous concerns about their impact on landscape and designated areas, but had felt unable to make their siting a policy matter, on the assumption that as this is dealt with under the Electricity Act, it is not an appropriate matter for an LP. The council has come to a more recent understanding, however, that "anything which has an effect on land use but not necessarily coming into the Planning Acts themselves can go into the local plan" (MDC Planning Officer A 2001).

8.8.2 Neighbourhood amenity and safety

The OHL policy is concerned, secondly, with the protection of residential and recreational areas from the effects of OHLs. Proposed changes to the deposit draft specified that effects on "amenity" are particularly in mind.

This aspect of the policy has been advanced within the context of new housing developments in proximity to existing OHLs, particularly on the edges of Frome, where OHLs skirt and enter one side of the town. Housing has recently been built on an infill site that is occupied by an OHL; in granting consent, the council negotiated a development-free corridor along the length of the OHL, in the interests of residential amenity (Figure 8.7). A second, nearby site is similarly constrained by a number of OHLs converging on a substation; the council's non-allocation of the site for housing, partly because of the presence of OHLs, has been the subject of an objection by a developer. These cases indicate that the inter-relationship of residential development and OHLs is an issue of current relevance and contention (MDC Planning Officer B 2001). This issue is also likely to be of future importance as the SP housing allocation for the district increases pressure for development on the edges of towns (Somerset County Council 1997).

Figure 8.6 400 kV OHL crossing the Somerset Levels, Mendip

Figure 8.7 Open space corridor along a 132 kV OHL, Frome, Mendip

Although the OHL policy seeks to limit the construction of new OHLs close to existing residential areas, rather than to control development near existing OHLs, the policy was "trying to make the argument" of not having housing in close proximity to OHLs and was formulated partly with the above two sites in mind (MDC Planning Officer A 2001):

The proposal for the first of the above sites generated some local concern:

> ... the local people speaking through the various town forums ... were saying to us, we're concerned about you building houses directly under power lines, because of the potential hazard, because it's not going to provide a good living environment for people (MDC Planning Officer A 2001).

Part of the public concern that was raised in relation to this case was anxiety about the perceived health effects of EMF. There have been particularly strong views expressed in the district about the possible effects of EMF, especially in relation to telecommunications equipment (CPRE Mendip Group Former Member 2001). For example, a proposal to construct a mobile phone mast in Frome was abandoned after local protests (Frome Town Council Town Clerk 2001). There has been considerable elected member interest in this issue, headed by one member following personal experience of illness, which he attributes to EMF exposure; he has a related campaigning role within his national party. Moreover, the council has recently lobbied government on the matter (MDC Elected Member 2001). Although these anxieties have focused on telecommunications equipment, there appears to have been some mutual reinforcement of concerns over these installations and OHLs (MDC Planning Officer A 2001). Anxieties about the possible effects of EMF, in general, have also been provoked by relevant media reports (MDC Planning Officer C 2001).

The council did not mention the possible health effects of EMF in the plan, either in the OHL policy or in an equivalent policy on the siting of telecommunications equipment. This is because the council did not feel that there is sufficient government support for this position (MDC Planning Officer A 2001). However, the plan does attempt to establish that EMF concerns could be relevant to planning by means of the hazardous installations policies: "we've put a policy in place which allows when it's demonstrated, if it can be demonstrated, that there is a health risk caused by power lines, we can use that policy in that context" (MDC Planning Officer C 2001). The plan therefore seeks a means of expressing EMF concerns while recognising the limitations imposed by scientific uncertainties and by the government's stance.

8.8.3 Public environmental awareness

Both of the above areas of policy, protection of landscape, and so on, and neighbourhood amenity have involved response to public concerns about OHLs and related issues. Although these concerns have not been expressed strongly in relation to OHLs, they have provided support for the council's policy position. They reflect a wider environmental awareness, in line with the social character of the area.

> It's the sort of area that's attracted a lot of people looking for an alternative lifestyle, we've got Glastonbury for instance ... but it's also a very self-sufficient area, so there's a very strong community feeling in each of the towns, who are each looking for the best for their towns as it were; then you've got things like the pressure groups, you've got CPRE who are quite strong. So it's coming from all sorts of different angles, from a traditional looking-after-the-countryside lobby ... then you've got the alternative greenies, to put it that way, who are meeting on a lot of these issues (MDC Planning Officer A 2001).

The council's approach to plan-making has placed strong emphasis on the involvement of local people; for example, early consultation took the form of seeking the public's views on what the future plan should propose rather than producing a written draft (MDC Planning Officer C 2001). The OHL and hazardous installations policies are in line with this approach in that they demonstrate responsiveness to the perceived views of the public on the environmental effects of HVDT installations.

8.9 CONCLUSION

MDC is an authority that has placed a high priority on environmental concerns, reflecting a culture of environmental awareness, both in the council and among the public. The importance that the council has placed on public participation in the preparation of its LP has enabled these concerns to be given particular weight in the formation of policy. The plan's handling of HVDT installations is consistent with this overall approach.

Within this context, certain factors have been instrumental in the formation of policy. With regard to the OHL policy's concern for landscape, and so on:

- the extent and quality of the district's designated sites and areas, along with the plan's strategy of conserving natural assets, have made the protection of landscape and associated features a priority;
- the existing, major OHL running through the area has drawn attention to the visual impact of OHLs in the district;
- concern about the effects of OHLs in the countryside expressed by interest groups, especially CPRE, has urged officers to make a policy response;
- the perceived possibility of new lines associated with wind turbines has given impetus to policy.

With regard to the OHL policy's concern for amenity and the hazardous installations policies' concern for health:

- new housing close to an OHL on the edge of Frome and questions over a nearby site have brought the issue of OHLs in residential areas into the local planning arena;
- local protests about the possible health effects of EMF, related primarily to telecommunications equipment, have urged the council to seek the protection of health;
- awareness of the government's stance on EMF has diverted policy on possible health effects to an anticipatory, conditional approach.

Finally, a realisation that LPs may cover any land-use planning topic has encouraged the inclusion of OHL policy. The components of policy formation by MDC are depicted in Table 8.2. The inter-relation of the principal components and the progression of policy through the DPP are summarised in Figure 8.8.

Table 8.2 Components of policy formation by Mendip District Council

BROAD CATEGORIES		WITHIN-CASE COMPONENTS	CROSS-CASE COMPONENTS*
INFLUENCING FEATURES	GEOGRAPHICAL, ECONOMIC AND SOCIAL FEATURES	Media reports on EMF research	Media reports on EMF research
		Structure plan housing allocation	Regional development pressure
		Extensive designated areas and sites	Extensive designated areas and sites
		Extensive rural areas	Extensive rural areas
		Culture of environmental awareness	Heightened public environmental awareness
	HVDT INSTALLATIONS (AND TELECOMMUNICATIONS)	Major OHL through countryside, affecting designated areas	Prominent installations in countryside
			Installations affecting designated areas/sites
		OHLs on edges of Frome	Installations near proposed development
		Perceived possibility of OHLs associated with wind turbines	Perceived likelihood of future HVDT projects
		Strong public and member opposition to telecommunications equipment	Local opposition to telecommunications equipment
	POLICY FRAMEWORK	Awareness of government stance on EMF	NRPB advice
	PLANNING EXPERIENCE	Development control cases involving OHLs in Frome	Electricity installation development control cases
	PLAN THEMES	Sustainable development	Sustainability
		Environmental capital approach	Protection of environmental features
		Public involvement in plan-making	Public participation
ACTORS	PLANNING OFFICERS	Knowledge of difficulties in referring to EMF	Caution against inclusion of EMF health concerns
		Importance given to perceived public concerns regarding OHLs	Importance attached to perceived public concerns
	ELECTED MEMBERS	Member-led telecommunications-related EMF campaign, arising from personal experience	EMF health/amenity concerns
			EMF-related personal anecdotal experience
	INTEREST GROUPS	CPRE landscape concerns	Landscape, etc. protection concerns
	PUBLIC	Broad public landscape concerns	Landscape, etc. protection concerns
		Telecommunications-related EMF protests	EMF health/amenity concerns
	ENVIRONMENTAL BODIES	Seeking to extend policy	Supporting aspects of policy
	ELECTRICITY INDUSTRY	Opposing proposed policy	Opposing aspects of policy
DPP	SCOPE OF PLAN	Broader understanding of plan content	Broadening of plan content
	REPRESENTATIONS	Countryside Agency seeking to extend policy	Supporting HVDT-related policy
		Electricity industry opposing proposed policy	Opposing HVDT-related policy

BROAD CATEGORIES		WITHIN-CASE COMPONENTS	CROSS-CASE COMPONENTS*
POLICY CONTENT	BROAD THEMES	Protection of designated areas	Countryside protection
		Protection of residential areas	Residential area protection
	EFFECTS OF HVDT INSTALLATIONS ON:	Visual quality of landscape	Visual quality: countryside
		Nationally and locally designated areas	Designated areas: national
			Designated areas: local
		Nature conservation areas	Ecological interests
		Archaeological sites	Archaeological sites
		Historical sites	Historical sites
		Amenity	Amenity
		Health	Public health (EMF)
		Safety	Public safety
		Residential areas	Residential, etc. areas
		Recreational areas	Recreational areas
	MEASURES TO OFF-SET EFFECTS	Opposition to new OHLs	Opposition to new HVDT installations
		Undergrounding of lines	Undergrounding of lines
		Distancing hazardous installations from existing development	Distancing new OHLs from existing development
		Distancing new development from existing hazardous installations	Distancing new development from existing OHLs

*These refer to components defined for the case studies as a whole; see Section 9.2.1

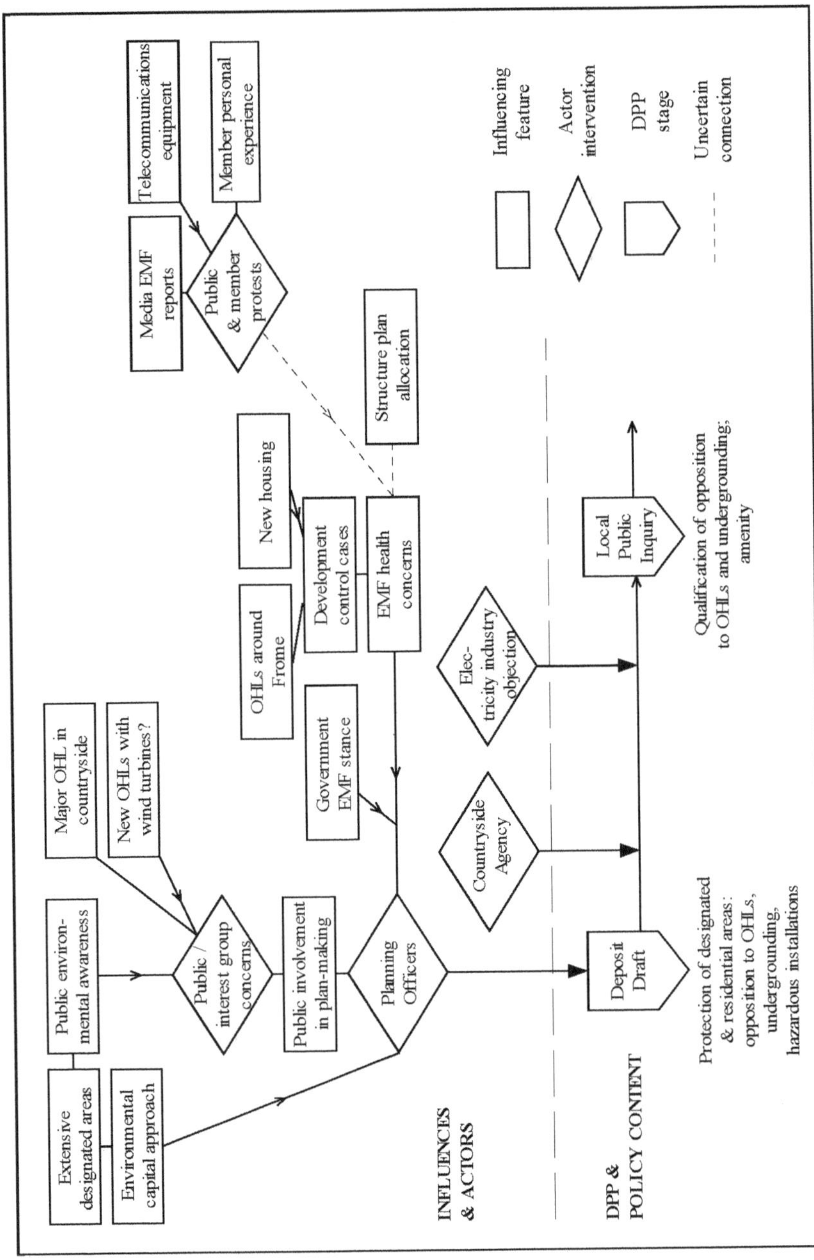

Figure 8.8 The process of policy formation by Mendip District Council

AYLESBURY VALE DISTRICT COUNCIL

8.10 INTRODUCTION

Aylesbury Vale District Council (AVDC) is a district authority in Buckinghamshire, southeast England. Central to the district is the town of Aylesbury, which is set in rural lowland, a "classically English landscape" (AVDC 1998). The authority is currently preparing its first district-wide LP , although precursor plans for Aylesbury and the "rural areas" provided a basis for this process. This case study was undertaken during the public local inquiry into the plan.

The deposit draft of the district-wide plan (AVDC 1998) reflects intense pressure for development in the region and is responding to the county SP's identification of the district as an area of growth. Aylesbury itself has experienced considerable recent expansion and is described in the plan as the main focus for the SP allocation of 8000 additional new homes by 2011 (Buckinghamshire County Council 1996). Four "Major Development Areas" are defined on the outskirts of the town, of a scale, which, the plan says, "is unprecedented for the town". Alongside this provision, however, is a strong sense of protection for the countryside and heritage.

The plan states sustainability and accessibility to be its two main themes. Sustainability is seen as including the minimisation of the consumption of natural resources; accessibility is understood in a broad sense to include the redressing of disadvantage.

8.11 ARTICULATION OF POLICY

The development of policies on HVDT installations is represented in Figure 8.9. Two former plans have contributed to the present formation of policy:

1. The Aylesbury Local Plan (AVDC 1991), covering the town, included a development constraint policy which restricted residential and much recreational development below existing OHLs; the boundary to which the restriction applied was the zone of maximum swing of the conductors.
2. The Aylesbury Vale (Rural Areas) Local Plan (AVDC 1995), covering the rest of the district, included a countryside protection policy resisting new OHLs in certain areas, especially special landscape areas. The justification given was the visually intrusive nature of OHLs; the policy also stated that OHLs should take account of landscape features. This plan also included the town plan's development constraint policy.

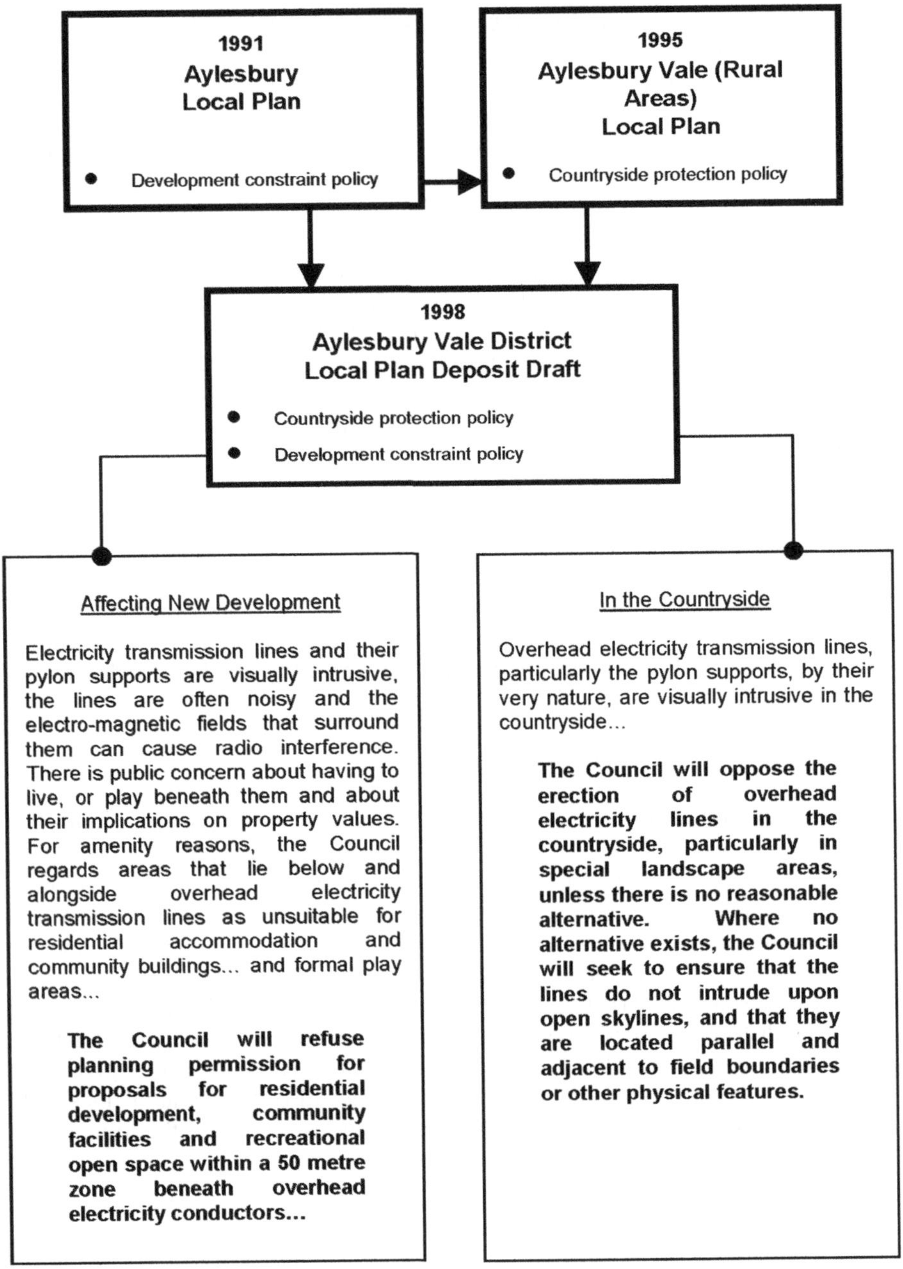

Figure 8.9 Policy development by Aylesbury Vale District Council

The deposit draft of the district-wide plan (AVDC 1998) took forward these policies:

1. The countryside protection policy was repeated, with the elaboration that pylons were intrusive "by their very nature".
2. The development constraint policy was considerably developed. Justification was given for the constraint, with reference to a number of environmental effects of OHLs, seen as giving rise to public concern. These were summed up as "amenity reasons" and provided the basis for a corridor of fixed width along the length of OHLs within which certain development would not be permitted.

Several representations were made in respect of these proposed policies. The countryside protection policy was supported by a number of bodies, including the Environment Agency, and was objected to by East Midlands Electricity. But much more significant representations were made in relation to the development constraint policy. Objections were received from house-building associations, East Midlands Electricity and NG. House builders claimed that the policy could diminish the development potential of certain sites. The electricity industry stated that the effects of OHLs on house prices are not within the remit of an LP; they also opposed the implication of health risk resulting from EMF. The council responded by insisting that its concerns fell within the context of social amenity; they argued that the absence of a corridor could lead to lower cost housing being built under OHLs, which could be seen as socially divisive. They also disputed the assumption that the policy appealed to health concerns, maintaining that it was based on amenity grounds alone.

On the other hand, some support was received locally for this policy; a local amenity society and an elected member acting in a private capacity called for the establishment of a corridor wider than the specified 50 m.

Some of these arguments were played out further at the plan inquiry. NG asserted that the principle of a fixed corridor was inflexible and unworkable. They argued that the degree of visual intrusiveness of OHLs depends on a number of factors, such as topography, tree cover and the design of nearby housing. A more helpful approach than setting a corridor of fixed width would be to treat each case on its merits.

In response, the council argued as follows:

- Other examples can be found in development control where fixed distances are used and accepted.
- A local precedent had been set for the policy; a 40 m corridor had been established alongside a 400 kV OHL as part of a new housing development at Coldharbour Farm. In this case, the corridor was established as a planning condition (Figure 8.10); it was seen as a successful design and was explicitly stated to have informed the proposed policy. The policy increased the width of the corridor from 40 m to 50 m on the basis of subjective judgement that this increase in distance leads to a substantial improvement in amenity.

- Loss of amenity for housing too close to OHLs is due to an accumulation of effects: visual intrusion, noise, vibration and interference with radio, television, and so on, possibly resulting in disparity in house prices and consequent social issues (although these "issues" were left undefined).

Figure 8.10 Corridor alongside a 400 kV OHL, Coldharbour Farm, Aylesbury

8.12 KEY POLICY ISSUES

8.12.1 Preservation of landscape

The countryside protection policy is concerned with the visual effects of OHLs in rural areas. Existing OHLs are prominent across wide parts of the district; for example, two 400 kV lines traverse the entire district, intersecting at a major substation. The presence of existing OHLs is therefore likely to have increased sensitivity to their visual effects in the open countryside.

The area is described as attractive and deserving of protection in the rural areas plan where the policy first appeared and in the district-wide plan. For example, the district-wide plan describes the countryside as "valuable for its natural beauty, its peace and quiet, its historical and cultural associations..." and speaks of the "need to protect the countryside for its own sake" (AVDC 1998). The plan then states its intention to exercise strict development control in the rural areas and to protect the landscape quality of the locally designated special landscape areas. A strong countryside protection theme therefore runs through the current plan.

This sense of need to preserve the rural landscape should be seen within the setting of strong development pressure on the district, due to its location between Milton Keynes to the north and Greater London to the south. Outward pressure from London is particularly acute, as most of the district lies just beyond two areas of severe development restraint, the Chilterns AONB and the Metropolitan Green Belt and is also served by good rail and road links to London. Both the rural areas and the district-wide plans place their countryside protection measures within this context. Although OHLs do not result directly from the growth of urban areas, the policy seeking to protect the countryside from OHLs can be seen as an expression of this wider concern to defend the character of the district against the "threat" of development.

The countryside protection policy also insists that all possible options should be considered before permitting their construction:

> We want to know if there are alternatives. Electricity companies have only thought about undergrounding. We consider a completely different route as an alternative; also whether electricity needs to be supplied, or whether an existing line can be upgraded. We would ask the electricity companies to demonstrate that they've looked at these alternatives (AVDC Planning Officer 2000).

This demonstrates a conviction that differing options exist in relation to HVDT installations and expresses unwillingness to accept the inevitability of OHLs. It also acts as a challenge to the electricity industry to give greater consideration to a wider set of options – or at least to communicate the reasoning behind their planning choices more openly.

8.12.2 Establishment of amenity zones

The development constraint policy seeks to protect the amenity of future residents from the potentially adverse effects of OHLs close to developed areas. This should be seen in the context of the pattern of OHLs in and around Aylesbury; a number of OHLs run alongside and towards the town, coming into close proximity with existing housing. The inter-relation of OHLs and housing is "something we've always had to consider" (AVDC Planning Officer 2000). For instance, this issue had been raised in the context of a planning appeal in 1989; the council then refused an application for new housing below an existing OHL on the grounds of harm to the amenity of potential residents. At that time, the inspector allowed the appeal, judging that the situation of housing underneath the OHL would not be materially different from that of existing nearby housing.

The presence of OHLs is, therefore, a factor in the design of future housing, and so on, in the district; for example, one of the major development areas covers an area with an existing concentration of OHLs. Setting a strategy for this issue is a priority, which the plan seeks to address; a fixed-width corridor free of certain development is offered as a solution. In fact, the site policy that defines the major development area mentioned above specifically provides for a 50 m zone beneath OHLs.

The formal expression of concerns about OHLs was made under the heading of amenity. Although specific effects of OHLs are referred to in the plan, it is more their overall effect and oppressive nature that are seen as a threat to amenity.

Local feelings of unease about OHLs were channelled through some members, and the council took on the role of defending the living conditions of both present and future residents:

> People have told us that they don't want to live under electric lines, therefore we shouldn't be building houses there; we don't build next to noxious factories, we try to avoid motorways, that's perfectly acceptable; we've taken the same stance with lines, we don't think that's unreasonable (AVDC Planning Officer 2000).

The preservation of amenity is seen here in terms of social equity. The council's argument about the danger of lower property values close to OHLs was an insistence that standards of amenity should not be apportioned on the basis of income (in line with the plan's overall theme of "accessibility"). This defence of potential residents' quality of living conditions took on a determined nature; elected members showed considerable interest in the policy as it was advanced, and officers committed considerable time and ingenuity to its defence at the public inquiry stage, making particular use of information available via the Internet. This combination of both political and professional commitment ensured that the policy was pursued to the fullest possible extent. It is likely that, even if the policy is not adopted, the local authority will still seek to ensure that no development takes place within 50 m of OHLs by means of development control powers (AVDC Planning Officer 2000).

The concerns mentioned specifically exclude any reference to the possibility of health risks arising from EMF, in contrast to policies advanced by some other LPAs. However, most of these authorities have not succeeded in this approach, following objections from the electricity industry, which has argued that the weight of scientific evidence does not support the likelihood of health risk resulting from EMF exposure (as, e.g. at Rotherham: Section 6.8.1). AVDC planning officers were aware of the direction of these scientific findings and also of the difficulties faced by LPAs attempting to base policies on health risk arising from exposure to EMF. They explicitly sought to avoid a similar outcome by basing their policy on other concerns and adopting the term "amenity zone"; the council's stance at the inquiry stage was along these lines.

Nonetheless, it is evident that health fears about EMF did provide some impetus for the policy. In particular, concern was aroused among some local residents and one elected member by reports of research into possible EMF effects obtained from the press and over the Internet. The question had initially been raised by a member following the experience of a local farmer whose land is affected by OHLs and who had lost one of his livestock to leukaemia (AVDC Elected Member 2001). This same member called for a corridor wider than 50 m; acting in a private capacity, he defended his position at the public local inquiry with explicit reference to EMF health concerns. Moreover, some of the support received for the development constraint policy was based on an assumption that it would provide some protection to health, despite its lack of reference to this (AVDC Planning Officer 2000).

The amenity zone strategy is effectively seen as a mitigating measure against the effects of OHLs. The insistence on a fixed distance, rather than allowing more flexibility in the layout of housing and so on, was encouraged by the local precedent set by the Coldharbour Farm scheme. Although in this case a corridor was established as a compensatory measure rather than out of concern for the effects of OHLs, it provided the council with a tangible example of how analogous planning difficulties might best be dealt with. Interestingly, the housing development along the edge of this corridor is designed to accentuate the form of the corridor itself (Figure 8.10), further strengthening the idea of well-defined limits to development.

Finally, the council's strategy of arguing for amenity zones without explicit reference to EMF concerns is seen as an example for other authorities to follow:

> ... we've had a lot of interest from other authorities who are aware that we're doing this, and with many authorities whose policies fell by the wayside as a result of National Grid representations. Some phoned us, via the grapevine ... (AVDC Planning Officer 2000).

The council therefore sees itself as attempting a new approach in relation to the co-existence of OHLs and residential development, one that takes account of the failure of other LPAs to establish their sought-after planning criteria.

8.13 CONCLUSION

The two policies referring to HVDT installations advanced by AVDC are intended for different settings: One aims to protect countryside from the visual intrusion of new OHLs, the other aims to maintain the amenity of future residents, primarily in and around Aylesbury, from the effects of existing OHLs. Both of these policies can be understood in the context of the favourable location for development in which Aylesbury Vale finds itself, in that they both seek to introduce ways of reducing adverse consequences of growth. The countryside protection policy falls within a range of policies seeking to defend the countryside against unrestricted development, and aims to be a preventative measure against the effects of one particular type of infrastructure. The development constraint policy aims to alleviate the co-existence of OHLs and major developments on greenfield sites. The policies can therefore be interpreted as the strategy of a concerned local authority to safeguard satisfactory conditions for its locality while under intense pressure for development.

Other features have also played a part in policy formation:

- OHL concern has an established history in the district;
- the density and location of existing OHLs in the district have increased sensitivity to their perceived effects and have motivated opposition to OHLs;
- media and Internet reports on the alleged health effects of EMF have aroused concerns that have increased commitment to policies;
- awareness of the experience of other LPAs in attempting to advance similar polices has informed planning officers about the limitations of policy;

- the scheme at Coldharbour Farm has provided policy makers with a model for amenity zones;
- the conviction and commitment of individual policy makers have driven the formation and defence of policies.

The policies presented by the council may also be interpreted as a progression in the formation of policies relating to HVDT installations as compared to those of other LPAs in that they aim to build upon the experiences of those other authorities. The components of policy formation by AVDC are depicted in Table 8.3. The inter-relation of the principal components and the progression of policy through the DPP are summarised in Figure 8.11.

Table 8.3 Components of policy formation by Aylesbury Vale District Council

BROAD CATEGORIES		WITHIN-CASE COMPONENTS	CROSS-CASE COMPONENTS*
INFLUENCING FEATURES	GEOGRAPHICAL, ECONOMIC AND SOCIAL FEATURES	Media and Internet EMF reports	Media reports on EMF research
		Southeast England development pressure; structure plan housing allocation	Regional development pressure
		Major development areas	Current major housing development
		Extensive rural areas	Extensive rural areas
	HVDT INSTALLATIONS (AND TELECOMMUNICATIONS)	High density of OHLs in countryside and around Aylesbury	High density of installations
			Prominent installations in countryside
			Installations near existing development
		OHLs across Major development area	Installations near proposed development
	PLANNING EXPERIENCE	History of OHL policy concern	History of HVDT-related policy
		Fixed-width corridor at Coldharbour Farm estate	Electricity installation development control cases
			Local precedents of measures to offset effects of HVDT
		Awareness of other authorities' difficulties in relation to EMF policies	Knowledge of other LPAs' experience
	PLAN THEMES	Sustainability	Sustainability
		Countryside protection	Protection of environmental features
		Accessibility	Social equity
ACTORS	PLANNING OFFICERS	Personal commitment to policy	Strongly supportive of policy
		Awareness of EMF research and difficulties faced by other authorities	Study of EMF research
			Caution against inclusion of EMF health concerns
		Response to public concerns	Importance attached to perceived public concerns
		Contact with authorities whose policies "failed"	Communication with officers from other authorities
	ELECTED MEMBERS	Strong support for officers	Strongly supportive of policy
		EMF health concerns	EMF health/amenity concerns

	BROAD CATEGORIES	WITHIN-CASE COMPONENTS	CROSS-CASE COMPONENTS*
		EMF-related anecdotal experience	EMF-related personal anecdotal experience
	INTEREST GROUPS	Amenity group supporting amenity zones	Supportive of/promoting policy
	PUBLIC	EMF health concerns	EMF health/amenity concerns
	ENVIRONMENTAL BODIES	Environment Agency supporting countryside protection policy	Supporting aspects of policy
	ELECTRICITY IN-DUSTRY	NG and East Midlands Electricity opposing policies	Opposing aspects of policy
	HOUSE-BUILDING INDUSTRY	House-building associations opposing development constraint policy	Opposing aspects of policy
DPP	REPRESENTATIONS	Amenity group, member and Environment Agency support	Supporting HVDT-related policy
		House building and electricity industries objections	Opposing HVDT-related policy
	LPA DEFENCE	Strong LPA defence	Strongly defending policy against representations
	Local public inquiry (LPI) / Examination in public (EiP)	Contestation	Contestation
POLICY CONTENT	BROAD THEMES	Countryside protection	Countryside protection
		Development constraint	Residential area protection
	EFFECTS OF HVDT INSTALLATIONS ON:	Visual quality: countryside	Visual quality: countryside
		Visual quality: residential areas	Visual quality: urban areas
		Special landscape areas	Designated areas: local
		Levels of public concern	Levels of public concern
		Amenity	Amenity
		Residential, etc. areas	Residential, etc. areas
		Property values	Property values
		Play areas	Recreational areas
		Noise levels	Noise levels
		Radio signals (EMF)	Radio signals, etc. (EMF)
	MEASURES TO OFFSET EFFECTS	Opposition to new OHLs	Opposition to new HVDT installations
		Careful routeing of new OHLs	Careful routeing of new OHLs
		Fixed-width zones along existing OHLs	Distancing new development from existing OHLs

*These refer to components defined for the case studies as a whole; see Section 9.2.1.

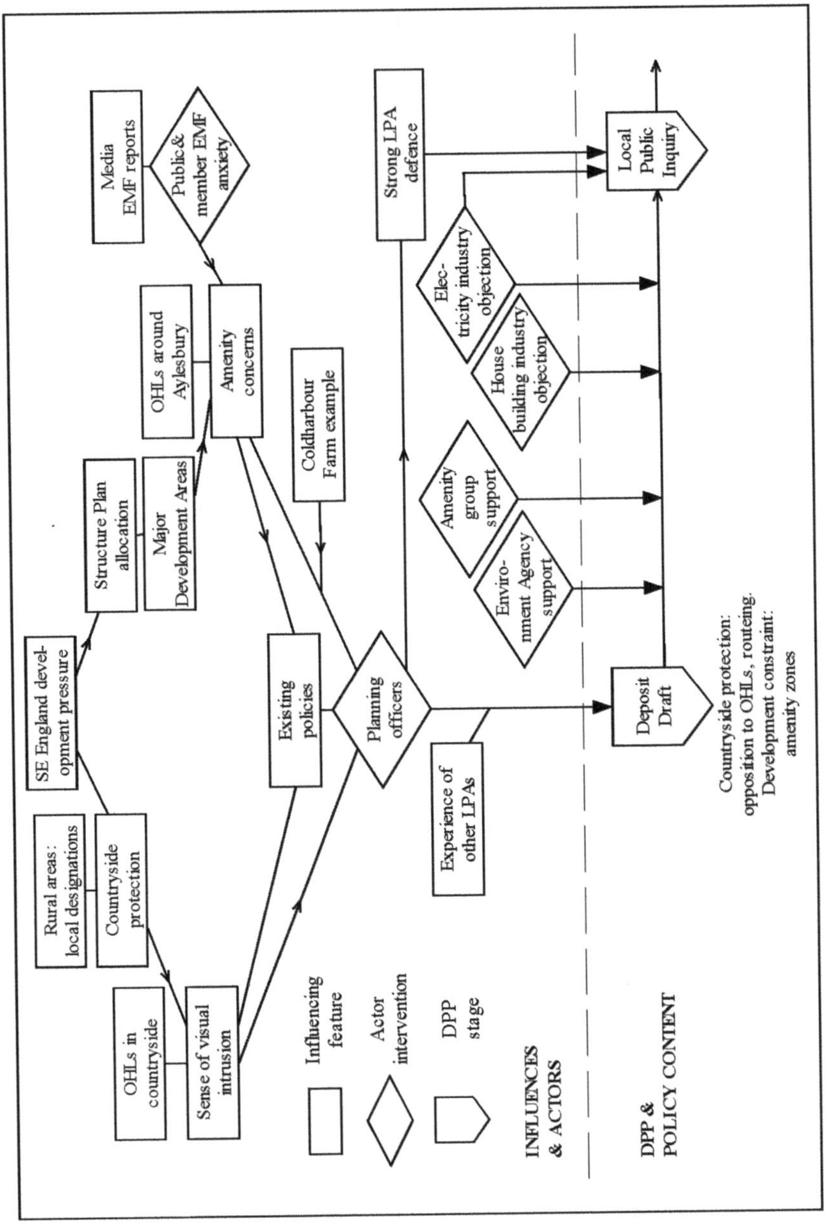

Figure 8.11 The process of policy formation by Aylesbury Vale District Council

BRIDGEND COUNTY BOROUGH COUNCIL

8.14 INTRODUCTION

Bridgend County Borough Council (BCBC) is a unitary authority in South Wales. The county borough reaches inland from the Bristol Channel, includes the town of Bridgend and rises onto the southeast Wales Plateau. Bridgend sits within the "M4 corridor" and has undergone recent economic and housing growth. This is in contrast to the upland part of the county borough, which contains former coal mining communities experiencing depopulation and increasing levels of social deprivation. The county borough contains diverse areas of coast and countryside, parts of which carry various landscape, conservation and other designations.

The council is currently preparing its first unitary development plan (UDP). This follows on from local government reorganisation in 1996, when BCBC was established as a unitary authority. Its area coincides with most of the former district of Ogwr, for which the Ogwr Borough Local Plan had been prepared under the Mid Glamorgan Replacement Structure Plan; these remain in force pending adoption of the UDP. References made below to "the plan" refer to the UDP. This case study was undertaken following publication of the deposit draft of the plan.

The deposit draft of the plan (BCBC 2001) takes sustainable development as its guiding principle and places emphasis on the protection of environmental features; this is set within the context of projections of population increase and major housing allocations in parts of the county borough.

8.15 ARTICULATION OF POLICY

The plan refers to high-voltage OHLs in two policies, in a chapter on "Energy and Utilities" in part 2 of the plan:
A utility services policy expresses concern that overground "service lines", including OHLs, may result in environmental damage; visual impact is mentioned. The need for careful line routeing is stressed. A preference for placing utility service lines underground is stated, although the difficulties of this for high-voltage lines are recognised; undergrounding is nonetheless expected in conservation areas. Careful routeing of lines is also considered necessary to avoid the sterilisation of mineral reserves.

1. A policy on OHLs and residential development seeks to maintain distance between OHLs and housing. It asserts that new OHLs will not be permitted across existing or proposed housing areas and that new housing will not be permitted beneath or adjacent to OHLs, for several reasons:

- OHLs can be visually dominant; although the design of residential developments can offset this, distancing houses from OHLs would facilitate the future undergrounding of those lines;
- maintenance of OHLs close to housing could be disruptive to residents;

- property values could be affected by having housing close to OHLs, owing to the visual effects of OHLs, or from public anxieties about the possible health risks of EMF exposure; on this last point, the plan states that there is no convincing evidence of such a health risk, but that the precautionary principle is applicable.

In addition, part 1 of the plan includes a general policy on energy and utilities that seeks to protect designated areas and "the local environment" from adverse effects of utility developments. Finally, a reference is made to energy installations in the "environment" section of the plan; this states the need to control various forms of development, including "power generation and its distribution network", in the locally designated landscape conservation areas. The main points of the above references are shown in Figure 8.12.

1997
Mid Glamorgan Replacement Structure Plan
- Utilities policy

2001
Bridgend CBC
UDP Deposit Draft

Part 1
- Utilities policy

Part 2
- Utility services policy
- HV lines and residential development policy

Energy and Utilities

Development for...utility services... will be favoured where it:
- **does not have an adverse effect on sites or areas of acknowledged international, national or regional importance;**
- **respects the overriding need to conserve the local environment...**
- **will not have an unacceptable visual, functional or physical impact either individually or cumulatively on its local and strategic settings**

...economic considerations, logistical interests, and matters related to the technology of the service and/or the location of the energy resource, should not be allowed to outweigh the need to conserve the environment as a whole.

Utility Services Development and the Environment

...the Council will expect the 'service companies' to route all above-ground service lines in a manner which will minimise damage or disturbance to the environment and reduce any potential sterilisation of identified mineral reserves... Where economically and technically feasible, and, where desirable on environmental grounds, services should be placed underground... The Council acknowledges, however, that... it would be unusual for major power lines to be placed underground except in exceptional circumstances...

High Voltage Power Lines and Residential Development

The construction of new high voltage power lines across existing, or proposed, residential development and/or proposals for new residential development beneath, or immediately adjacent to existing or proposed high voltage power lines will not be permitted.

High voltage power lines, their carrying towers, and other infrastructure can have a dominant visual impact on the landscape. This can, however, be mitigated in part by the design and layout of new residential development... If development is not permitted beneath, or adjacent to such lines, this allows easier scope for a future relocation of the line underground... Similarly, residential amenity can be disrupted by the regular maintenance of the line...

While the Council has noted ... that there is no convincing evidence to date of a causal link between exposure to electromagnetic fields and ill health, it recognises however, that there are genuinely held public perceptions to the contrary, which taken with the problems of visual amenity may affect land values / house prices. It will therefore continue to assess such proposals with due reference to the Precautionary Principle...

Figure 8.12 Policy development by Bridgend County Borough Council

These policies were introduced in a consultation draft of the plan (BCBC 2000). Representations were made on this draft by NG, who objected to

- the call for the undergrounding of lines, because of financial, technical, ecological and archaeological difficulties involved;
- the reference to the precautionary principle, asserting that there is inadequate evidence of health risk from EMF exposure;
- the resistance to constructing OHLs and housing in proximity to each other, stating that the impact of OHLs on nearby development can be more appropriately minimised by other means, such as careful line routeing and design of developments.

The council did not make any changes to the plan as a result of these representations; NG's objections were therefore reiterated following publication of the deposit draft.

In preparing the plan, the council had regard to the existing SP and LP. The Mid Glamorgan Replacement Structure Plan (BCBC 1997) contained a utilities policy, which provided a basis for the council's proposed policies. In particular, it called for the careful routeing of OHLs in order to minimise any impact on the environment or the sterilisation of mineral reserves, and for consideration to be given to the possibility of undergrounding lines where feasible. The Ogwr Borough Local Plan does not mention these specific concerns, but does raise the issue of the environmental effects of utility development, particularly in the context of their privatisation: "There is consequently a greater role for local planning to play ... in regulating the environmental consequences of the development aspirations of the privatised utility companies through the development plan" (Ogwr Borough Council 1995, p. 167).

8.16 KEY POLICY ISSUES

8.16.1 Visual effects of utility service provision

The plan's concern for the overall environmental effects of utility services, including OHLs, originated partly with the existing SP. Concern centres mainly on the visual impact of installations; with respect to OHLs, the SP makes reference to their possible visual effects on designated areas, "the landscape" and residential areas.

Some incursion of OHLs into areas of landscape value has been experienced owing to two OHLs crossing the county borough; these include a 400 kV line crossing hilly ground at the base of the coalfield plateau (Figure 8.13). However, these OHLs do not appear to be of major concern:

> I would say that most of them have been there for so long, people are used to seeing them and they don't tend to notice them ... The fact that it dives in and out of valleys ... it's not that noticeable (BCBC Planning Officer A 2001).

Figure 8.13 400kV OHL crossing hilly countryside, Ogmore Valley, Bridgend County

This area of policy appears to derive more from the plan's wider emphasis on restricting development in the countryside in order to preserve landscape quality; this emphasis is found in the "environment" section of the plan, where, as noted above, electricity installations are specifically mentioned. Most concern is expressed for the protection of designated landscape areas. These include the "strategic coalfield plateau and associated valley sides" defined by a South Wales–wide landscape assessment and which received SP attention; it incorporates much of the northern half of the county borough. The policies seeking to minimise the visual effects of OHLs are partly a reflection of the importance attached to this landscape area:

> Once you get above a certain height in southeast Wales, you stand on that plateau, and you miss the valleys in between and you just see basically a flat surface going right out; it's an old coastal plain ... As a result of that, the strategic policy is particularly important, because what you don't want to see is disruptive visual impact which can be seen across the whole of southeast Wales (BCBC Planning Officer A 2001).

The local importance attached to this and other landscape areas is also illustrated by the council's support for "Greening of the Valleys", a cross-agency initiative aiming at improving the environment of the coalfield plateau's valleys (BCBC 2001). The plan also makes reference to past industrial activities that have despoiled many sites and states that protecting the visual amenity of the county borough is essential for enhancing its attractiveness and improving the quality of life of its residents (BCBC 2001).

Concern is also expressed for the visual effects of OHLs upon residents; this is couched partly in terms of amenity and also in terms of landscape, the implication being that views from residential areas are significant. This is also illustrated by the plan's reservations about possible wind farm schemes; one reason given for resistance to them is that "most 'upland' wind farm proposals are likely to be visible from a settlement" (BCBC 2001).

This emphasis on the county borough's landscape quality is one aspect of the council's stress on sustainability. The council sees itself as having particular strengths in this regard as indicated, for example, by the decision to have an environmental appraisal of the plan carried out by outside consultants (BCBC Planning Officer A 2001).

The plan suggests that the visual effects of "service lines" can be offset by the careful routeing or, preferably, undergrounding of lines. This approach follows directly from the SP, which states the same objectives in similar wording. The plan acknowledges that undergrounding cannot usually be expected for high-voltage OHLs, but the measures suggested are nonetheless considered applicable in principle to all lines. The undergrounding of lines has been a sensitive issue locally, especially in the valleys, where electricity supply was historically provided overhead, on poles:

> ... over time, the people in those communities have seen the pits disappear, they've seen the eyesores disappear, like the buckets, the pithead gears and so forth, they've seen their telephones go underground, and ... the last thing they saw going underground in some places was the electricity cables. So there was always a very keen mood expressed over the last 20, 25 years in favour of undergrounding ... (BCBC Planning Officer A 2001).

The idea of undergrounding lines therefore appears to have been extended from smaller-scale lines to larger-scale OHLs. Local precedents have also been set for this policy notion and for that of careful line routeing; recent highway construction, following the opening of the M4 motorway, has necessitated the re-routeing and undergrounding of certain OHLs (BCBC Planning Officer A 2001).

8.16.2 Access to mineral reserves

The plan also advocates careful routeing of lines in the interests of preventing sterilisation of mineral reserves. This policy concern is inherited directly from the SP and reflects awareness of the importance of mineral exploitation in the county borough. Although the deep mining of coal has ceased, open cast coal mining now operates, and limestone extraction accounts for the largest concentration of quarrying in South Wales. This latter resource is used as an aggregate and as a fluxing stone for the nearby steelworks at Port Talbot. The continued supply of flux limestone from the county borough is considered to be essential to the economic viability of the steelworks (BCBC Planning Officer A 2001), and therefore of particular, even national, importance (BCBC 2001). The protection of potential mineral reserves from developments such as OHLs is therefore seen to be in the vital economic interests of the county borough and beyond.

8.16.3 Maintaining distance between high-voltage lines and housing

The policy dealing with the inter-relationship of OHLs and residential development was introduced into the plan without any precursor in the existing SP or LP. Although it deals with both the construction of new OHLs near existing housing and new housing near existing OHLs, it is the latter scenario that appears to be the council's primary planning concern in this regard.

This issue has become significant locally, following housing pressure associated with economic growth along the M4 corridor. The plan allows for over 7000 new dwellings in the county borough during the plan period, mostly in and around the town of Bridgend; development has been focused on a site known as *Broadlands*. This site is, however, bisected by a 275 kV OHL that was originally routed to skirt around the town. Following the initial allocation of the site, a development brief was prepared, which stipulated that no housing should be built within 20 m of the centre of the OHL (Ogwr Borough Council 1996). Subsequent development has taken account of the presence of the OHL by locating open space and roadways below and alongside it; some of the spaces created between housing are of considerable width (Figure 8.14). Similarly, a much smaller site has involved residential development in the vicinity of an existing lower-voltage OHL; in this case, houses have been built in closer proximity to the line (Figure 8.15). The expansion of Bridgend across the paths of existing OHLs has therefore brought to the fore an issue of land-use conflict for which the council has sought practical solutions and is now seeking to establish a policy position.

Figure 8.14 Open space along a 275kV OHL, Broadlands housing estate, Bridgend

Figure 8.15 Housing in proximity to a 132 kV OHL, Bridgend

Despite the Broadlands Development Brief's fixed-width zone within which no housing is permitted, the UDP does not specify a minimum distance to be maintained between housing and OHLs. The council now prefers to allow for flexibility in the design of housing around existing OHLs and to use the space underneath and alongside OHLs for a variety of non-residential purposes:

> ...it can form better boundaries within developments as well, because you don't always want to have public open space beneath them... and trees can go into those areas... because quite obviously there's going to be some people who are not even going to like walking under them... So it's better to patchwork it, with various bits of open space and what have you, and see if you can get some roads underneath it here and there... (BCBC Planning Officer A 2001).

Of the reasons given in the plan for maintaining distance between OHLs and housing, concern about the possible health effects of EMF has been the most dominant. There had been officer awareness of this issue during preparation of the LP in the early 1990s, but this did not come to policy expression at the time. However, health concerns did lie behind the fixed-width zone set out in the Broadlands Development Brief, even though no reference is made to them in the document. Moreover, the issue was raised more forcibly, especially by elected members, during the early consultation stages of preparing the plan; this led to specific references being made to health fears and the precautionary principle (BCBC Planning Officers A and B 2001). An initial response from NG to proposed policy wording led officers to recommend removing references to EMF, but:

> Members were very, very adamant ...; they said, well even if there isn't any physical or medical risk that is proved as such, there is still a perceived risk, which is going to make life very difficult for such developments to go ahead underneath such powerlines, or indeed above, because the public perception is that it's not safe to live there (BCBC Planning Officer A 2001).

Member concern about the possible health effects of exposure to EMF had been raised largely by media reports covering the issue during the 1990s. Personal experience and anecdotal evidence have also contributed to anxieties; for instance, one member refers to the death from cancer of an acquaintance who had lived in close proximity to an electricity pylon (BCBC Elected Member A 2001). Similar concerns regarding EMF exposure have also been expressed in relation to telecommunications equipment; local protests against proposed equipment have been particularly vociferous in one part of the county borough (BCBC Elected Member B 2001), and the council has recently passed a resolution banning the location of mobile phone masts on council-owned school playing fields and children's play areas.

Another reason given for distancing housing from existing OHLs is to facilitate the possible future undergrounding of those lines. This objective is concerned mostly with the visual effects of OHLs, although there is a suggestion that undergrounding might also alleviate people's health fears:

> If a millionaire was found and they all went underground, you wouldn't have to dig up everybody's gardens and arrange easements, because you haven't got development under them to start with, and vice versa, if people are not living underneath them, they're not so worried about their effect, and as a result, properties sell better (BCBC Planning Officer A 2001).

The council has found some support for maintaining distance between housing and OHLs from the reluctance of house builders to take on sites close to OHLs; developers looking at Broadlands and elsewhere were avoiding sites close to OHLs for fear of having difficulty selling houses there. However, developers who have taken on these sites have shown a preference for building housing as close to OHLs as possible in order to maximise a site's potential (BCBC Planning Officer B 2001) and have challenged the retention of "cordons sanitaires" (BCBC Planning Officer A 2001). This runs counter to the council's view that the possible lower value of houses close to OHLs argues against their construction in the first place.

8.17 CONCLUSION

In preparing its first UDP, BCBC has inherited from the former two-tier plan system certain policy concerns relating to HVDT installations, which have provided an impetus for the current statement of policy. These concerns have been brought to the fore by a number of factors:

- designated landscape areas, especially those that have received regional attention and hold local importance, have motivated a strong landscape protection stance, coupled with the aim of improving damaged landscapes;
- the presence of existing OHLs across the county borough has added to the sense of need to protect landscape from further OHLs;
- local precedents have informed the policy notions of undergrounding and careful routeing of OHLs;
- the importance of mineral extraction to local industry has encouraged the protection of mineral reserves from OHLs;
- the plan's overall theme of sustainability has given further impetus to the formation of environmental protection policies.

In addition, the recent pressure for housing development arising from Bridgend's favourable location on the M4 corridor has resulted in the extension of residential areas up to and beyond an OHL that was originally routed to avoid the town. This has provoked an additional policy area dealing with the co-existence of housing and OHLs. A number of features have been instrumental in this respect:

- concerns among members and, to a lesser extent, officers about the alleged health effects of exposure to EMF have motivated the aim of maintaining distance between OHLs and housing; media coverage and anecdotal accounts have played a part in heightening these anxieties;
- the council considered the precautionary principle to be supportive;

- the council has taken the house-building industry to be supportive of its approach;
- the development brief prepared for Broadlands acted as a precursor for development plan policy on this point.

The components of policy formation by BCBC are depicted in Table 8.4. The inter-relation of the principal components and the progression of policy through the DPP are summarised in Figure 8.16.

Table 8.4 Components of policy formation by Bridgend County Borough Council

BROAD CATEGORIES		WITHIN-CASE COMPONENTS	CROSS-CASE COMPONENTS*
INFLUENCING FEATURES	GEOGRAPHICAL, ECONOMIC AND SOCIAL FEATURES	Media reports on EMF research	Media reports on EMF research
		M4 corridor development pressure	Regional development pressure
		Extensive rural area containing designated areas	Extensive designated areas and sites
			Extensive rural areas
		Industrial decline in valleys	Economic decline/poor environmental conditions
		Important mineral reserves	Important mineral reserves
		Broadlands estate	Current major housing development
	HVDT INSTALLATIONS (AND TELECOMMUNICATIONS)	OHLs crossing rural area	Prominent installations in countryside
		275 kV OHL through Broadlands development	Installations near proposed development
		Protests against proposed equipment	Local opposition to telecommunications equipment
	POLICY FRAMEWORK	Mid Glamorgan Structure Plan	Supportive regional/structure planning policy
		Ogwr Borough Local Plan, Broadlands Development Brief	Supportive local planning instruments
		Precautionary Principle	Precautionary Principle
	PLANNING EXPERIENCE	Inherited concerns from two-tier plan system	History of HVDT-related policy
		Broadlands and other development areas	Electricity installation development control cases
		Examples of undergrounding and careful line routeing	Local precedents of measures to offset effects of HVDT
	PLAN THEMES	Sustainable development; protection of environmental features	Sustainability
			Protection of environmental features
ACTORS	PLANNING OFFICERS	Strong planning officer involvement	Strongly supportive of policy
		Local value attributed to landscape	Landscape valuation/assessment
		Caution against inclusion of EMF health concerns	Caution against inclusion of EMF health concerns
	ELECTED MEMBERS	Strongly supportive of policy, especially EMF references	Strongly supportive of policy
		EMF health concerns	EMF health/amenity concerns

BROAD CATEGORIES		WITHIN-CASE COMPONENTS	CROSS-CASE COMPONENTS*
		EMF-related personal anecdotal experience	EMF-related personal anecdotal experience
	ELECTRICITY INDUSTRY	NG objections to policy	Opposing aspects of policy
DPP	REPRESENTATIONS	Objections from NG	Opposing HVDT-related policy
POLICY CONTENT	BROAD THEMES	Environmental protection	Environmental/countryside protection
		Residential area protection	Residential area protection
	EFFECTS OF HVDT INSTALLATIONS ON:	Visual quality of landscape and residential areas	Visual quality: countryside
			Visual quality: urban areas
		National and local designated areas	Designated areas: national
			Designated areas: local
		Public EMF concerns	Levels of public concern
		Visual amenity	Amenity
		EMF health concerns	Public health (EMF)
		Residential areas	Residential, etc. areas
		Land values/house prices	Property values
		Access to mineral reserves	Access to mineral reserves
POLICY CONTENT	MEASURES TO OFFSET EFFECTS	Statutory duties of "service companies"	Reliance on statutory responsibilities of utilities
		Opposition to new OHLs in residential areas	Opposition to new HVDT installations
		Careful routeing of new OHLs	Careful routeing of new OHLs
		Undergrounding of lines	Undergrounding of lines
		Distancing new OHLs from existing development	Distancing new OHLs from existing development
		Distancing new development from existing OHLs	Distancing new development from existing OHLs
		Design of residential development near OHLs	Sensitive design of development near OHLs

*These refer to components defined for the case studies as a whole; see Section 9.2.1.

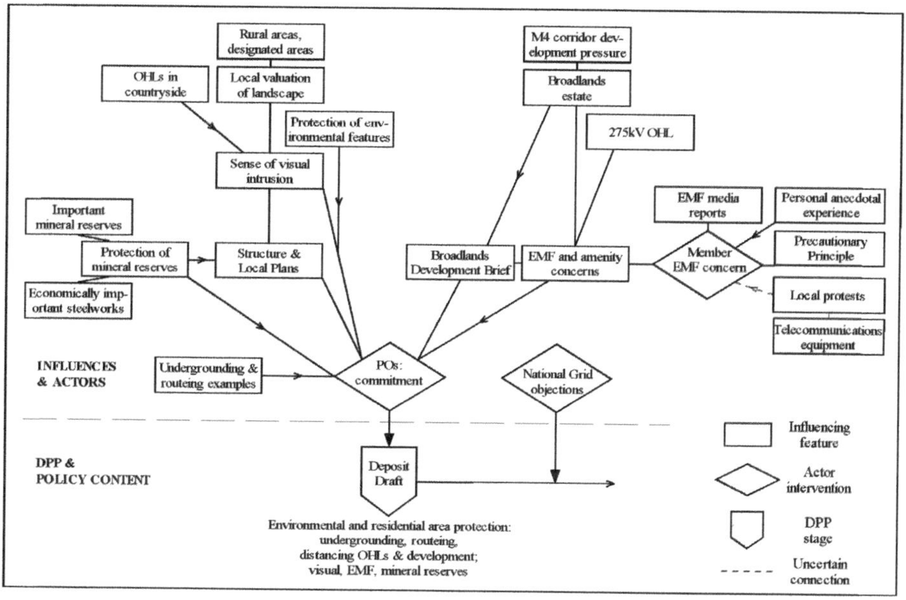

Figure 8.16 The process of policy formation by Bridgend County Borough Council

CHAPTER 9

Analysis of the Case Studies

9.1 INTRODUCTION

Chapters 6 to 8 have explored the processes of high-voltage distribution and transmission (HVDT)-related policy formation in 10 individual case studies. These processes were described in terms of the broad elements of policy formation set out in Figure 1.2: influencing features, actors, development plan process (DPP), policy content and the inter-relation between these. Analysing the formation of policy in terms of these elements allowed a "within-case" summary of each case study to be presented, by means of a table listing the components that were most instrumental in policy formation (e.g. Table 6.1), and a network diagram portraying the most important links between those components and the progression of the policy through the DPP (e.g. Figure 6.4).

This chapter draws from the individual case studies the key components and linkages that are at work in the HVDT-related policy formation, particularly those which occur across a number of local planning authority (LPA)s. This enables a more generalised picture of policy formation to be made, which emphasises the policy processes operating across different authorities. A "cross-case" analysis is thus presented, which draws upon data from the individual case studies. This analysis is mostly based upon the within-case summary tables and network diagrams, with reference to the complete case study accounts where necessary.

Individual case studies are abbreviated below and in subsequent chapters as follows. In tables, the abbreviations defined in Chapters 6 to 8 are used, whereas in the text, place names are used:

Case Study	Tables	Text
Swindon Borough Council	SBC	Swindon
Rotherham Metropolitan Borough Council	RMBC	Rotherham
Newham Council	NC	Newham
Tynedale Council	TC	Tynedale
Norfolk County Council	NCC	Norfolk
Snowdonia National Park Authority	SNPA	Snowdonia
Redcar & Cleveland Borough Council	RCBC	Redcar & Cleveland
Mendip District Council	MDC	Mendip
Aylesbury Vale District Council	AVDC	Aylesbury Vale
Bridgend County Borough Council	BCBC	Bridgend

The final section of this chapter presents the results of a brief survey of a number of LPAs that have not expressed HVDT-related policy concern. This provides a different perspective on the priorities involved in policy formation to that provided by the case studies (Section 4.4.5).

9.2 CROSS-CASE ANALYSIS

9.2.1 Synthesis of the components of policy formation

The within-case tables list the components that have been most instrumental in the policy formation for each individual case study. They show the features, actor involvement and others specific to each case, and are described as within-case components; these are shown, for example, in the second main column of Table 6.1. However, these tables also show these elements translated into "cross-case" components, seen, for example, in the third main column of Table 6.1. Cross-case components are found to be in common across the 10 case studies, and were defined to allow a synthesis to be carried out of the factors at work throughout the case studies.

These cross-case components were defined by finding similarities among the within-case components for all 10 case studies, such that elements for different case studies could reasonably be grouped together. A single heading was then devised for the group of components. For example, Norfolk's "favourable location for energy projects" and Snowdonia's "renewable and other energy resources" are both translated into the common heading "potential for energy generation projects".

The complete list of the cross-case components occurring throughout all the case studies is shown in Table 9.1, in the second main column. This table also shows, in matrix form, which of these components occurs in each case study. It therefore portrays a synthesis of factors operating across the 10 case studies, and provides a basis for tracing patterns of policy formation among them.

Table 9.1 Components of policy formation by the 10 case study authorities

	BROAD CATEGORIES	CROSS CASE COMPONENTS	SBC	RMBC	NC	TC	NCC	SNPA	RCBC	MDC	AVDC	BCBC
INFLUENCING FEATURES	GEOGRAPHICAL, ECONOMIC AND SOCIAL FEATURES	Privatisation / liberalisation of electricity industry							●			
		Pressure for energy projects					●	●				
		Media reports on EMF research	●	●	●					●	●	●
		Regional development pressure	●							●	●	●
		Current major housing development	●								●	●
		Extensive designated areas and sites				●	●	●	●	●	●	●
		Extensive rural areas				●	●	●	●	●	●	●
		Proximity of national park							●			●
		Economic decline/poor environmental conditions		●	●				●			●
		Concentration of hazardous industries							●			
		Important mineral reserves & associated industry										●
		Potential for major development			●							
		Potential for tourism				●						
		Potential for energy generation projects					●	●	●			
		Heightened public environmental awareness	●							●		
	HVDT INSTALLATIONS (AND TELECOMMUNICATIONS)	High density of installations		●					●		●	
		Prominent installations in countryside				●	●	●	●	●	●	●
		Installations affecting designated areas/sites				●	●	●	●	●		
		Installations near existing development	●	●	●				●		●	
		Installations near proposed development	●	●	●					●	●	●
		Recent/current HVDT projects				●	●		●			
		Perceived likelihood of future HVDT projects					●	●	●	●		
		Local opposition to telecommunications equipment				●				●		●

	BROAD CATEGORIES	CROSS CASE COMPONENTS	SBC	RMBC	NC	TC	NCC	SNPA	RCBC	MDC	AVDC	BCBC
INFLUENCING FEATURES	POLICY FRAMEWORK	Supportive national planning policy				●		●				
		Supportive regional/structure planning policy		●	●	●			●			●
		Supportive local policy instruments	●		●		●		●			●
		Supportive HVDT-related public inquiry decisions			●				●			
		Comparable neighbouring authority policy		●					●			
		Precautionary Principle	●	●	●							
		International precedents		●								
		NRPB advice	●	●	●				●	●		
		Electricity industry guidelines		●								
		National committee/association support						●				
	PLANNING EXPERIENCE	History of HVDT-related policy			●			●	●		●	●
		Recent/current HVDT projects				●	●		●			
		HVDT-related public inquiries			●				●			
		Electricity installation development control cases		●		●				●	●	●
		Local precedents of measures to offset effects					●	●	●		●	●
		Knowledge of other LPAs' experience	●	●		●			●		●	
	PLAN THEMES	Sustainability					●	●		●	●	●
		Protection of environmental features	●			●	●	●	●	●	●	●
		National park purposes						●				
		Regeneration/environmental improvement		●	●				●			
		Social equity									●	
		Public health		●								
		Public participation								●		
	INSTITUTIONS	New institutional arrangements			●			●				

BROAD CATEGORIES	CROSS CASE COMPONENTS	SBC	RMBC	NC	TC	NCC	SNPA	RCBC	MDC	AVDC	BCBC
PLANNING OFFICERS	Strongly supportive of policy	●		●	●		●	●		●	●
	Landscape valuation/assessment				●	●	●	●			●
	Study of EMF research		●		●					●	
	Caution against inclusion of EMF health concerns		●	●				●	●	●	●
	Importance attached to perceived public concerns								●	●	
	Communication with officers from other authorities		●							●	
ELECTED MEMBERS	Strongly supportive of policy		●	●		●		●		●	●
	Opposition to HVDT projects					●					
	Landscape and other protection concerns					●		●			
	EMF health/amenity concerns		●	●		●		●	●	●	●
	EMF-related personal anecdotal experience								●	●	●
	Communication with other authorities		●								
INTEREST GROUPS	Opposition to HVDT projects				●		●	●			
	Supportive of/promoting policy	●								●	
	Landscape and other protection concerns						●	●	●		
	Study of EMF research	●									
PUBLIC FORUM	Public forum advancing policy ideas	●									
PUBLIC	Opposition to HVDT projects				●		●	●			
	Landscape and other protection concerns					●	●	●	●		
	EMF health/amenity concerns	●	●	●		●		●	●	●	
ENVIRONMENTAL BODIES	Supporting aspects of policy		●		●				●	●	
DEVELOPMENT CORPORATION	Supporting aspects of policy			●							

	BROAD CATEGORIES	CROSS CASE COMPONENTS	SBC	RMBC	NC	TC	NCC	SNPA	RCBC	MDC	AVDC	BCBC
	ELECTRICITY INDUSTRY	Opposing aspects of policy	●	●	●	●	●	●	●	●	●	●
	HOUSE-BUILDING INDUSTRY	Opposing aspects of policy	●	●							●	
DPPs	SCOPE OF PLAN	Broadening of plan content								●		
	REPRESENT-ATIONS	Supporting HVDT-related policy	●	●		●				●	●	
		Opposing HVDT-related policy		●	●	●		●	●		●	
	Local public inquiry (LPI) / Examination in public (EiP)	Strongly defending policy against objections		●	●	●			●	●		●
	LPA DEFENCE	Contestation		●	●	●			●		●	
		Inspector/Panel recommended alterations	●	●	●	●	●	●	●			
HVDT-RELATED POLICY CONTENT	BROAD THEMES	Countryside protection				●	●	●	●	●	●	●
		Residential area protection	●	●	●	●			●	●	●	●
		Encouraging regeneration			●							
	EFFECTS OF HVDT INSTALLATIONS ON:	Visual quality: countryside				●	●	●	●	●	●	●
		Visual quality: urban areas		●	●	●			●		●	●
		Designated areas: national				●	●	●	●	●		●
		Designated areas: local				●	●		●	●	●	●
		Ecological interests				●	●		●	●		
		Archaeological sites				●	●		●	●		
		Historical sites				●				●		
		Levels of public concern	●	●							●	●
		Amenity	●	●			●			●	●	●
		Public health (EMF)	●	●	●	●			●	●		●
		Public safety	●	●		●			●	●		
		Residential and other areas	●	●	●	●			●	●	●	●

	BROAD CATEGORIES	CROSS CASE COMPONENTS	SBC	RMBC	NC	TC	NCC	SNPA	RCBC	MDC	AVDC	BCBC
		Property values									●	●
		Recreational areas			●	●	●		●	●	●	
		Noise levels		●							●	
		Radio signals and others (EMF)									●	
		Land use and regeneration efforts			●							
		Access to mineral reserves										●
HVDT-RELATED POLICY CONTENT	MEASURES TO OFFSET EFFECTS	Reliance on statutory responsibilities of utilities		●								●
		Opposition to new HVDT installations			●	●			●	●	●	●
		Removal/re-routeing of existing OHLs		●	●			●				
		Careful routeing of new OHLs			●	●		●	●		●	●
		Undergrounding of lines		●	●	●	●	●	●	●		●
		Distancing new OHLs from existing development				●				●		●
		Distancing new development from existing OHLs	●	●	●					●	●	●
		Sensitive design of development near OHLs			●							●
		Siting new energy schemes near existing OHLs					●					

9.2.2 The broad themes of the HVDT-related policy

In the analysis of DPPs across England and Wales, two main areas of policy interest were revealed within HVDT-related policy, dealing with the effects of HVDT installations in countryside and urban settings respectively (Section 5.6.3); these were found to be overlapping, yet distinct areas of concern. Similarly, Table 9.1 indicates that different policy themes are operating within the case studies; these are shown as shaded rows in the final section of the table. Three broad themes occur, the first two frequently, and the third in an isolated case:

- countryside protection (CP) (occurring in seven case studies);
- residential area protection (RAP) (eight case studies);
- encouraging regeneration (one case study).

The first two of these themes, "countryside protection" and "residential area protection", echo the rural and urban foci of the policy areas identified in the analysis of DPPs across England and Wales. The cross-case and DPP analyses therefore corroborate each other in identifying two principal, broad areas in HVDT-related policy. Moreover, the individual case study accounts suggest that LPAs often have in mind policy aims approximating to the policy themes defined here. Aylesbury Vale, for example, pursued two separate policies, one opposing new overhead line (OHL)s in the countryside, the other prohibiting residential development close to existing OHLs. This cross-case analysis therefore takes the policy themes as the basis for organising the factors revealed in the case studies; this, in turn provides a means of further investigating the themes themselves.

9.2.3 Cross-case analysis by policy themes

From Table 9.1, those components of policy formation which relate to each of the three broad policy themes have been identified, and then transferred to separate tables. Table 9.2, for example, lists the components that shape the CP theme. This table also shows, in matrix form, the occurrence of these components in each of the seven case studies concerned.

Table 9.2 Components shaping HVDT-related policy on "countryside protection"

COUNTRYSIDE PROTECTION	TC	NCC	SNPA	RCBC	MDC	AVDC	BCBC
INFLUENCING FEATURES							
Extensive designated/rural areas	●	●	●	●	●	●	●
Prominent HVDT installations affecting country-side/designated areas	●	●	●	●	●	●	●
Recent/current/possible HVDT projects	●	●	●	●	●		
Energy market pressures and generation potential		●	●	●			

COUNTRYSIDE PROTECTION		TC	NCC	SNPA	RCBC	MDC	AVDC	BCBC
External pressures on countryside (development/ national park)					●		●	
Economic interests in countryside (tourism/minerals)		●						●
Plan strategy of sustainability/environmental protection		●	●	●	●	●	●	●
Plan aim to counter poor environmental conditions					●			●
Public environmental awareness and participation in plan-making						●		
LPA history of HVDT-related policy				●	●		●	●
Local HVDT-related public inquiries/development control cases		●			●			
Local precedents of measures to offset the effects of HVDT			●	●	●			●
Supportive policy/planning decisions at national/regional level		●		●	●			●
Supportive local policy instruments/neighbouring authority policy			●		●			
Support from national bodies and reformed institutional arrangements				●				
ACTORS & DPPs								
Broadening of plan content						●		
Planning officers: strongly supportive of policy		●		●	●		●	●
Planning officers: landscape valuation/consideration of public concern		●	●	●	●	●		●
Elected members: strongly supportive of policy			●		●			
Elected members: opposition to/concerns about effects of HVDT			●		●			
Interest groups/public: opposition to/concerns about effects of HVDT		●	●	●	●	●		
Environmental bodies: supporting aspects of policy		●				●	●	
Electricity industry: opposing aspects of policy		●	●	●	●	●	●	●
Strong LPA defence of policy against objections		●		●	●			
Inspector/panel recommended alterations		●	●	●	●			
POLICY CONTENT								
Effects of HVDT installations on:	Visual quality of the countryside	●	●	●	●	●	●	●
	Ecological/archaeological/historic interests	●	●		●	●		
	Amenity/recreation		●					
	Designated areas: national	●	●	●	●	●		●
	Designated areas: local	●	●		●	●	●	●
	Access to mineral reserves							●
Measures to offset effects	Reliance on statutory responsibilities of utilities							●
	Opposition to new HVDT installations	●		●	●	●	●	

COUNTRYSIDE PROTECTION		TC	NCC	SNPA	RCBC	MDC	AVDC	BCBC
	Removal/re-routeing of existing OHLs			●				
	Careful routeing of new OHLs			●	●		●	●
	Undergrounding of lines	●	●	●	●	●		●
	Siting new energy schemes near existing OHLs		●					

However, it should be noted that Table 9.2 represents a progression of data from Table 9.1, as further generalisation of data was sought at this point: similar components were merged and given more generic names. Table 9.2 therefore presents, for the CP theme, more generic cross-case components than are shown in Table 9.1. Similarly, Tables 9.3 and 9.4 were drawn up for the other two themes (Table 9.4 differs from the other two tables in that it is based upon only one case study: Section 9.3.3.)

Table 9.3 Components shaping HVDT-related policy on "residential area protection"

RESIDENTIAL AREA PROTECTION	SBC	RMBC	NC	TC	RCBC	MDC	AVDC	BCBC
INFLUENCING FEATURES								
HVDT installations near existing/proposed residential development	●	●	●		●	●	●	●
Current HVDT project; perceived likelihood of future projects					●			
Local electricity installation development control cases		●				●	●	●
Local precedents of measures to offset effects of HVDT							●	●
Development pressure, major housing development	●						●	●
Media reports on EMF research	●	●	●			●	●	●
Local opposition to telecommunications equipment				●		●		●
Public environmental awareness/participation in plan-making	●					●		
LPA history of HVDT-related policy							●	
Plan emphasises public health/social equity/environmental improvement		●			●		●	
Supportive regional planning guidance		●						
Supportive local policy instruments/neighbouring authority policy	●	●			●			●
Appeal to Precautionary Principle/international precedents	●	●	●					●
Reference to NRPB advice/electricity industry guidelines	●	●	●		●	●		

RESIDENTIAL AREA PROTECTION		SBC	RMBC	NC	TC	RCBC	MDC	AVDC	BCBC
ACTORS & DPPs									
Broadening of plan content							●		
Planning officers: strongly supportive of policy		●			●			●	
Planning officers: study of EMF research			●					●	
Planning officers: contact with other LPAs/caution against EMF reference			●	●		●	●	●	●
Planning officers: importance attached to perceived public concerns							●	●	
Elected members: strongly supportive of policy			●	●		●		●	●
Elected members: EMF-related personal anecdotal experience							●	●	●
Interest groups: supportive of/promoting policy		●	●					●	
Public forum: advancing policy ideas		●							
Public: EMF concerns		●	●	●		●	●	●	
Electricity industry: opposing aspects of policy		●	●	●	●			●	●
House-building industry: opposing aspects of policy		●	●					●	
Strong LPA defence of policy against objections			●					●	
Inspector-recommended alterations		●	●			●			
POLICY CONTENT									
Effects of HVDT Installations on:	Visual quality in urban areas		●	●	●	●		●	●
	Public health (EMF) and safety	●	●	●	●	●	●		●
	Amenity	●	●				●	●	●
	Noise levels/radio signals, etc.		●					●	
	Levels of public concern regarding health risks, etc.	●	●					●	●
	Property values							●	●
Measures to offset effects	Reliance on statutory responsibilities of utilities		●						
	Opposition to new HVDT installations				●	●	●		●
	Removal/re-routeing/careful routeing of OHLs		●		●				
	Undergrounding of lines		●		●		●		
	Distancing development from OHLs	●	●	●	●		●	●	●
	Sensitive design of development near OHLs			●					●

Table 9.4 Components shaping HVDT-related policy on "encouraging regeneration"

ENCOURAGING REGENERATION		NC
INFLUENCING FEATURES		
Major regeneration strategy in area of industrial decline and environmental blight		●
Major OHLs near proposed development		●
Supportive regional and local planning policy and inquiry decisions		●
History of HVDT-related policy		●
ACTORS & DPPs		
Planning officers: strongly supportive of policy		●
Elected members: strongly supportive of policy		●
Development corporation: strongly supportive of policy		●
Electricity industry: opposing aspects of policy		●
Strong LPA defence of policy against objections		●
Inspector-recommended alterations		●
POLICY CONTENT		
Effects of HVDT installations on:	Visual quality in urban areas	●
	Land use	●
Measures to offset effects	Removal/re-routeing/careful routeing of OHLs	●
	Undergrounding of lines	●
	Opposition to new HVDT installations	●

Although these tables list the components at work in each of the policy areas, they do not indicate the inter-relation between the components. To show this, a network diagram was drawn up for each theme, as was carried out for the individual case studies. Figure 9.1 shows the diagram for the CP theme. This was based upon the network diagrams for the seven case studies that include the CP theme (Figure 7.3 and others). Within-case components shown in these diagrams were translated into the cross-case components, and frequently occurring linkages between these components were identified. Thus, a generalised picture was drawn up of components and linkages occurring in the seven case studies. Moreover, the number of case studies in which different components and linkages occur is represented in the diagram by varying thickness of boxes and lines. Similarly, Figures 9.2 and 9.3 were drawn up for the other two policy themes (Figure 9.3 differs from the other two diagrams in that it is based upon only one case study; see below.)

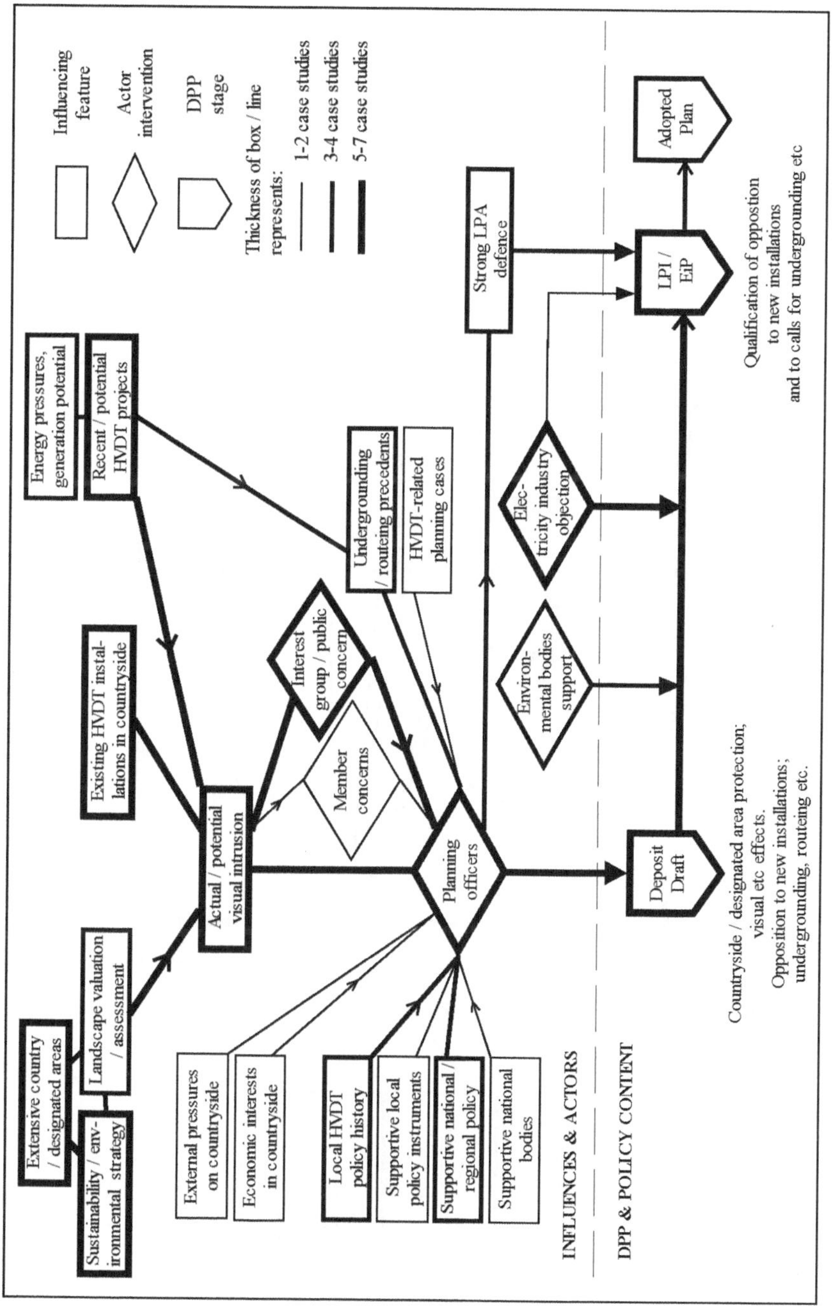

Figure 9.1 Generalised process of policy formation on "countryside protection"

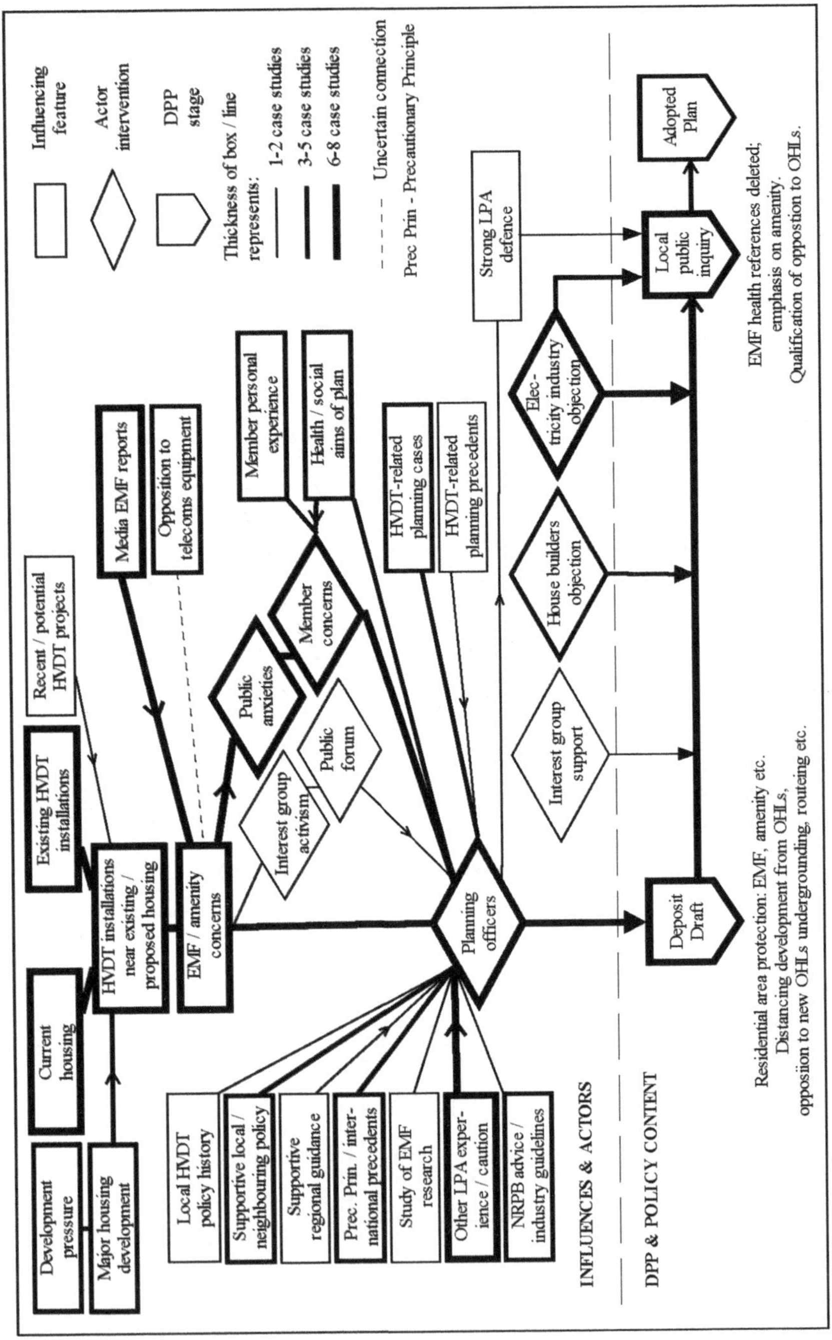

Figure 9.2 Generalised process of policy formation on "residential area protection"

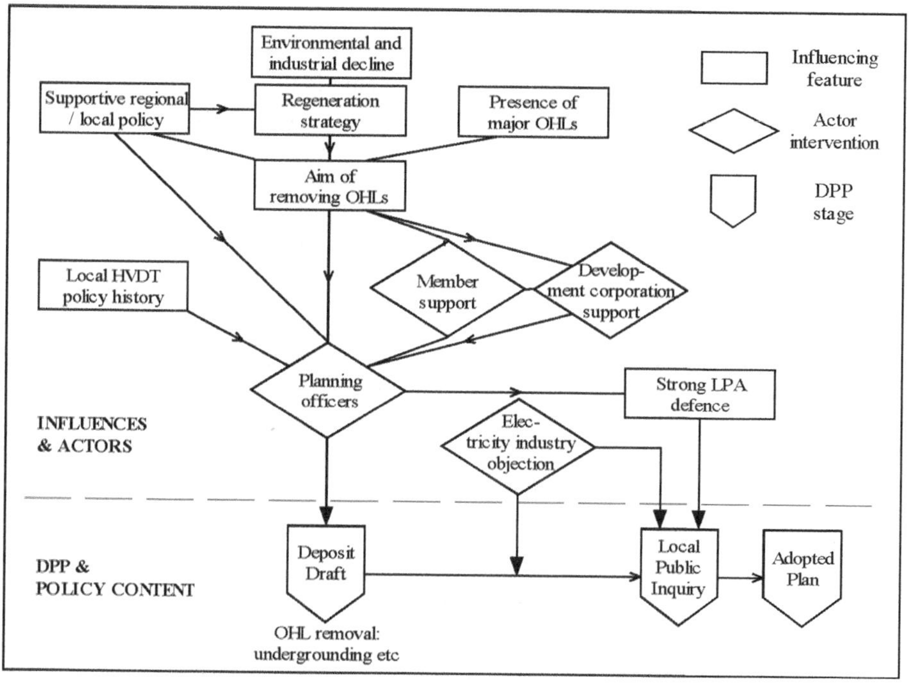

Figure 9.3 Generalised process of policy formation on "encouraging regeneration"

9.3 POLICY THEMES

9.3.1 Countryside protection

Table 9.2 and Figure 9.1 summarise the CP policy theme as expressed in the case studies. All the seven authorities dealing with this policy theme cover rural or semi-rural areas; the three authorities covering urbanised areas are not included.

This policy theme is concerned in all cases with the visual quality of the countryside; other effects of HVDT installations on natural and historical features are also frequently raised. These concerns are focused upon the protection of designated areas. Isolated concern is also expressed for recreational areas and access to mineral reserves. These aims are typified by Tynedale, which includes an area of outstanding national beauty (AONB), locally designated landscape areas and sites of archaeological interest among those it seeks to protect.

A universal feature of these case studies is the presence of extensive rural and designated areas. This is closely linked to a stance of sustainability, environmental protection, or environmental enhancement being taken for the plans as a whole. Some cases also demonstrate a strong sense of value being attributed to the local landscape, sometimes related to formal landscape assessment. For example, Norfolk shows links between its sustainable development strategy, its extensive designated areas and local appreciation of the landscape.

Another universal feature of this policy theme is the presence of prominent HVDT installations affecting rural/designated areas. Also frequent is the recent occurrence of local HVDT projects, or the prospect of such projects in the foreseeable future. This latter scenario is sometimes linked to pressures for new energy generation schemes. These factors are again exemplified by Norfolk, which already has intrusive HVDT installations in its area, and faces the possibility of future installations in connection with conventional and renewable power projects.

This policy theme, therefore, results from the convergence of a sense of highly valued rural and designated areas, and a sense of intrusion upon them from existing HVDT installations, heightened in some cases by recent or impending HVDT projects. In addition, it is linked in some cases to a wider need to protect the countryside from external pressures, such as development pressure, or to other strategic aims, such as the promotion of tourism.

These perspectives are especially strong among planning officers; public or interest group concerns or opposition to the effects of HVDT installations in the countryside are also a common factor. Snowdonia, for example, has a history of strong opposition from amenity groups and the wider public to the effects of energy installations. Planning officers are central to policy formation; they are aware of the issues raised by interest groups, the public and the elected members; in Mendip, they were particularly sensitive to public concerns, because of an explicit emphasis on public participation in plan-making. Elected member concern is marked in only two authorities.

Most of the case study authorities have been recently involved in HVDT-related consent procedures, either as consultees or as the consenting authority. This has involved the officers directly in considering the effects of HVDT installations. Recent projects have also provided local precedents for the undergrounding and re-routeing of lines. In Redcar & Cleveland, for example, proposals for the "North Yorkshire" line involved the LPA in public inquiries and in negotiations for line routeing; a precedent for undergrounding was also set by this line. While drafting a policy, officers have sometimes built upon the existing development plan policy; they have also had recourse to national or regional policy guidance, referring either generally to CP measures, or explicitly to HVDT installations. Policy precedents and backing have also been found in other local policy instruments, the policies of neighbouring authorities and the national bodies.

A range of measures is proposed by the authorities to offset the effects of HVDT installations in the countryside, such as the undergrounding of lines and the careful routeing of new OHLs. Most authorities mention two or three such measures; Bridgend, for instance, refers to the environmental responsibilities of utility companies, routeing lines in a manner that will minimise damage, and the possibility of undergrounding of lines in exceptional circumstances.

Policies proposed in deposit draft plans have met with support in some cases from environmental bodies, such as English Nature, but opposition in every case from the electricity industry. Objections have led some LPAs to defend their proposed policies strongly; changes to proposed policy have been recommended by an inspector or a panel, typically leading to LPAs qualifying their opposition to new HVDT installations and their demands for lines to be placed underground.

9.3.2 Residential area protection

Table 9.3 and Figure 9.2 summarise the RAP policy theme as expressed in the case studies. The eight case study authorities dealing with this policy theme cover urbanised and semi-rural areas, and one rural area.

This policy area is concerned with the possible effects of HVDT installations in urban settings, especially in residential and recreational areas. This is nearly always connected with the alleged health effects of exposure to electromagnetic field (EMF); other issues, such as visual quality and amenity, are also frequently raised, and occasionally noise and the possible lowering of property values, and so on. These concerns are typified by Rotherham, which raises the issues of health, safety, amenity and growing public concern and possible implications of OHLs for housing, schools, hotels and so on.

The most common feature of these case studies is the presence of HVDT installations in close proximity to residential development. This is not generally due to the construction of new HVDT installations in urban areas; in most cases, established housing is already close to installations. Sometimes, however, new housing is planned close to HVDT installations because of regional development pressure. Bridgend, for example, lying on the M4 corridor, has allocated an area bisected by an OHL as a major housing site.

Another frequent feature of this policy theme is the influence of media reports on the possible effects of EMF exposure on human health; these have contributed to public anxieties regarding the proximity of housing to HVDT installations. Elected members have often voiced these concerns. In Newham, for instance, press reports in the early 1990s about the suspected effects of EMF generated strong feeling among people living near a major OHL, culminating in public meetings and raising strong member support. A number of additional features may have added to these concerns; local opposition to telecommunications equipment, also a source of EMF, may be reinforcing wider concern about EMF (in Mendip, for example). Members have been motivated by personal anecdotal accounts of the possible effects of EMF. Wider concerns for health and welfare have also shaped member attitudes, as is illustrated by Rotherham's "health for all" strategy. In Swindon, strong activism by Friends of the Earth mobilised local anxieties about EMF.

Planning officers have been responsible for drafting the policy in response to public and member attitudes. In so doing, they have sought backing, from policy tools, such as the Precautionary Principle, international policy precedents and regional guidance. Officers have also looked to other local policy instruments, such as supplementary planning guidance and the policies of neighbouring authorities.

Sometimes, officers have made contact with other authorities who have sought to pursue similar policies, and they have undertaken some study of research into the possible effects of EMF exposure. They have also referred to the National Radiological Protection Board (NRPB) advice on the issue, usually following representations from the industry. Although, in some authorities, officers have been strongly supportive of the policy theme, the contacts and research that they have undertaken, in most cases, have led them to caution against pursuing the policy along the lines of health fears. Officers in Aylesbury Vale, for example, avoided any reference in their proposed policy to health risk.

The most frequent measure proposed to ensure the protection of residential areas is the maintaining of distance between urban development and OHLs; in some cases, a set distance is specified. Most commonly, policy seeks to prohibit new developments from being constructed in close proximity to existing OHLs, but in some cases, new OHLs would be opposed near existing developments. Other proposed measures include the undergrounding of lines, and the sensitive design of development near OHLs. While drafting a policy, officers in some authorities have drawn on specific local examples of development control that have dealt with the presence of HVDT installations, thus setting policy precedents. Aylesbury Vale, for instance, proposed a policy of prohibiting residential development within a 50-m zone along OHLs; this was based upon a specific instance of development control in a district in which a zone of this kind had been enforced.

Policies proposed in deposit draft plans have met with some support from interest and amenity groups, but also strong opposition from the electricity industry and, occasionally, house builders. Objections have resulted in most LPAs modifying their policies either before or at the inquiry stage of plan preparation, though LPAs have continued to strongly defend their proposed policies in some cases. Some changes to policies have been recommended by inspectors, usually involving the deletion of references to health, with emphasis being placed instead on amenity.

9.3.3 Encouraging regeneration

Table 9.4 and Figure 9.3 summarise the expression of the encouraging regeneration policy theme. The description of this theme differs from that of the previous two, in that it is based upon only one case study, Newham; a generalised picture of policy formation cannot therefore be drawn from multiple case studies, as has been done above. However, summarising the components relevant to encouraging regeneration, as revealed in the Newham case study, provides a presentation of the theme comparable with the CP and RAP themes.

Newham covers an urbanised area. Its aim of encouraging regeneration is concerned with the impact of HVDT installations on efforts to bring about economic regeneration, due to both their adverse visual effects, and the constraints on land use that they impose. The policy seeks the removal of OHLs, preferably by undergrounding lines, and opposes any future OHLs.

The features combining these goals are the overall strategy of regeneration, linked to the area's history of industrial decline, and the dominant presence of major OHLs. Both the regeneration strategy and the objective of removing OHLs receive strong regional and local policy support. There has also been unanimous support for the policy from officers, members and a development corporation. An HVDT-related planning case, in the form of a wayleave hearing, also provided some support for the policy initiative. However, objections to the draft policy were made by the National Grid. The LPA defended their approach strongly, though the inspector during plan inquiry recommended some qualification of the policy.

9.4 COMPARISON OF THE POLICY THEMES

Differentiating the policy themes as above, and investigating the influences and processes at work within them, have revealed patterns of policy formation particular to each theme. The differences between them can also be investigated by means of a comparative analysis. Table 9.5 presents the relative importance of cross-case components for the three different themes. The list of components shown in this table has been compiled from the previous tables, with a further degree of generalisation taking place. The table indicates, by symbols, the frequency of each component for each policy theme (or, in the case of "encouraging regeneration", whether a component occurs).

Table 9.5 Components shaping the three broad themes of HVDT-related policy

COMPONENTS SHAPING HVDT-RELATED POLICY		CP	RAP	ER
INFLUENCING FEATURES				
Geographical, economic & social features	Extensive designated/rural areas	■		
	Energy market pressures and generation potential	◘		
	Pressures on countryside, economic interests in countryside	⊡		
	Development pressure, major housing development		◘	
	Media reports on EMF research		■	
	Industrial decline and environmental blight			▪
HVDT installations (and telecommunications)	HVDT installations affecting areas upon which policy is focused	■	■	▪
	Recent/current/possible HVDT projects	◘	⊡	
	Local opposition to telecommunications equipment		◘	
Planning experience	Local HVDT-related public inquiries/development control cases	⊡	◘	
	Local precedents of measures to offset the effects of HVDT	◘	⊡	
Policy framework	Supportive national/regional guidance/planning decisions, etc.	◘	⊡	▪
	Supportive local policy instruments/neighbouring authority policy	⊡	◘	▪
	LPA history of HVDT-related policy	◘	⊡	▪
	Appeal to Precautionary Principle, etc., reference to NRPB advice, etc.		◘	
Plan themes	Supportive overall plan strategy	■	⊡	▪
	Public participation in plan-making	⊡	⊡	

COMPONENTS SHAPING HVDT-RELATED POLICY		CP	RAP	ER
ACTORS & DPPs				
Planning officers	Strongly supportive of policy	■	◘	▪
	Drawing on professional expertise (landscape/EMF)	◘	⊡	
	Contact with other LPAs		◘	
	Importance of perceived public concerns; broadening plan content	⊡	⊡	
Elected members	Strongly supportive of policy	⊡	◘	▪
	Active concern regarding/opposition to effects of HVDT	⊡	◘	
Interest groups	Supportive of/promoting policy	⊡	⊡	
	Active concern regarding/opposition to effects of HVDT	◘	⊡	
Public forum	Advancing policy ideas		⊡	
Public	Active concern regarding/opposition to effects of HVDT		■	
Environmental bodies	Supporting aspects of policy	◘		
Development corporation	Strongly supportive of policy			▪
Electricity industry	Opposing aspects of policy	■	■	▪
House-building industry	Opposing aspects of policy		◘	
LPA defence	Strongly defending policy against objections	◘	⊡	▪
Inspector / panel	Recommended alterations	◘	◘	▪
POLICY CONTENT				
Effects of HVDT installations on:	Visual quality of the countryside, designated areas	■		
	Ecological/archaeological/historic interests	◘		
	Recreation/access to mineral reserves	⊡		
	Visual quality in urban areas		■	▪
	Public health (EMF)		■	
	Public safety, public concern		◘	
	Property values/noise levels/radio signals, etc.		⊡	
	Amenity	⊡	◘	
	Land use			▪
Measures to offset effects	Reliance on statutory responsibilities of utilities	⊡	⊡	
	Opposition to new HVDT installations	◘	◘	▪
	Removal/re-routeing/careful routeing of OHLs	◘	⊡	▪
	Undergrounding of lines	■	◘	▪
	Siting new energy schemes near existing OHLs	⊡		
	Distancing development from OHLs		■	
	Sensitive design of development near OHLs		⊡	

■	All/nearly all cases	CP	Countryside protection
◘	Some cases	RAP	Residential area protection
⊡	Isolated cases	ER	Encouraging regeneration
▪	Newham		

9.4.1 Influencing features

The comparative importance of influencing features can be summarised from Table 9.5 as follows:

- *Geographical/economic and other features* are demonstrated in all three themes. Extensive designated/rural areas and energy generation pressures occur consistently in relation to CP, as do media reports on EMF issues and housing pressure for RAP. Although these features are of very different characters, they demonstrate the importance of background influences and external pressures in heightening the perceived need for protection. "Encouraging regeneration" (ER) also has a broader economic/environmental feature providing a background to the policy concerns.

- *Existing HVDT installations* are a universal feature for all three themes, especially installations that have a direct bearing upon the areas of policy concern. Recent/potential HVDT projects are also a common feature for CP, whereas opposition to the presence of telecommunications equipment is possibly a feature for RAP.
- *Local HVDT-related planning experience* is a feature for both CP and RAP. For RAP, this is more related to development control cases, where, for example, restrictions have been placed on housing development near OHLs. For CP, precedents arising out of direct experience of dealing with HVDT projects are of greater importance, such as a local case of undergrounding being carried out.
- *Policy framework* is important to all three themes. For CP, the national/regional framework is more significant, such as regional guidance on CP, whereas, for RAP, local policy instruments are more important, such as supplementary planning guidance on the design of residential areas. A history of local HVDT-related policy figures in all three themes, but is most important for CP. RAP also draws on additional guidance seen as having relevance to EMF issues, such as the Precautionary Principle. ER calls on policy support at several levels.
- *Plan strategies* that are relevant to policy are important to all three themes. This is especially so for CP, which relates to broader sustainability and environmental strategies, but also for RAP, calling on health strategies, for example. ER is based on an overall plan strategy of regeneration. Public participation in plan-making is also relevant to both CP and RAP in one case.

9.4.2 Actors and DPPs

The comparative role of the principal actors and their involvement in DPPs can be summarised from Table 9.5 as follows:

- *Planning officers* show support for all three themes, but most strongly for CP. Their professional expertise, on landscape valuation for instance, is also of most importance in relation to this theme. Officer studies of EMF issues are also of significance in isolated cases of RAP. Officer contact with other LPAs is only seen in RAP, where it occurs frequently, again in relation to EMF issues. There are isolated cases in both CP and RAP of officer initiative in broadening the content of plans, and in giving particular attention to public concerns.

- *Elected members* show strong support in all three policy themes, but this is most apparent in relation to RAP. Similarly, they demonstrate active interest in relation to RAP in a number of cases, reflecting their take-up of EMF concerns.
- *Interest groups* take the initiative in supporting or promoting a policy in only isolated cases for both CP and RAP, but they show a broader concern in a larger number of cases in relation to CP. This reflects the interests of amenity groups and others in landscape and related issues.
- *Active general public* concern features only in RAP, though very frequently; this reflects the expression of public EMF exposure anxieties.
- *Environmental bodies*, such as English Nature and Council for the Protection of Rural England (CPRE), express support for CP policy only, in a number of cases.
- *The electricity industry* opposes the policy in all three themes, in nearly all cases; this mainly reflects the intervention of the National Grid, but also of other representatives of the industry. *The house-building industry* also opposes the policy in a number of cases in relation to RAP, reflecting their interest in developments in the vicinity of HVDT installations.
- LPAs *defend their policies* strongly in relation to all three themes, though only in isolated cases in relation to RAP; this reflects their awareness of difficulties in referring to EMF-related health fears.
- *Inspector/panel* recommended changes occur in all three themes, though only in isolated cases in relation to RAP, reflecting the pattern of LPA policy defence.

9.4.3 Policy content

The comparative content of the policies can be summarised from Table 9.5 as follows:

- The various effects of HVDT installations with which policy is concerned divide clearly along the lines of the different policy themes. CP is concerned mostly with visual quality in the countryside, protection of the designated areas and ecological, archaeological and other interests. RAP is concerned mostly with visual quality in the urban areas, health and safety issues and levels of public concern. Each of these two themes also deals in isolated cases with other issues, such as, for RAP, effects on property values. The only issue in common between these two themes is amenity. ER deals with visual effects in the urban areas and land use.
- The proposed measures to offset the effects of HVDT are more commonly shared between the three themes. Opposition to HVDT installations, and undergrounding, are the most frequent; consideration of line routeing is also frequent, especially for CP. Other measures are proposed that are specific to either CP or RAP. These include, for RAP, maintaining distance between the development and OHLs; this is, moreover, the most frequently advanced measure for RAP.

9.5 "NO HVDT-RELATED POLICY" AUTHORITIES

The above analysis, along with that presented in Chapter 5, reveals strong trends at work in the formation of HVDT-related policy among many LPAs, particularly in certain areas of England and Wales. This raises the question of why other LPAs have not pursued this policy area, and whether there are factors actively dissuading the formation of the policy. To test for this possibility, a brief telephone survey was carried out of a number of "no HVDT-related policy" LPAs (as outlined in Section 4.4.5). Twenty LPAs were selected, representing two categories:

- LPAs that share characteristics similar to the 10 case study authorities, but have no advanced HVDT-related policy;
- LPAs located in parts of England and Wales with a relatively low level of policy concern (Figures 5.6–5.11).

Table 9.6 summarises the comments made by planning officers in these LPAs.

These comments show, that in most of these authorities, HVDT installations are not considered as a planning issue requiring policy attention. This is despite the occurrence of OHLs and other installations throughout most of the authorities' areas. In some cases, OHLs are thought of as established features in the landscape, and not excessively intrusive; these include the Newark & Sherwood District Council, which has a high concentration of OHLs and a power station in its area. In one case (South Holland District Council), the lack of policy interest was connected with a sense of the local landscape not being highly valued.

However, some of these authorities have raised HVDT-related concerns, especially in relation to the perceived EMF effects of OHLs, but have been reluctant to pursue the policy. For instance, Sheffield City Council decided against doing so, partly in the light of neighbouring Rotherham's difficulties over the same issue.

A number of officers did not see policy formation as an appropriate means of dealing with HVDT-related planning issues. One saw legislation governing the electricity industry as adequate in itself (York City Council), while others saw development control measures as more suitable, such as in relation to EMF concerns (Cotswold District Council). Another suggested reason for the non-inclusion of HVDT-related policy in development plans is the current advice to make the plans shorter and more focused (Mid Devon District Council).

More general points arising from these comments are:

- telecommunications equipment is now of greater policy concern than OHLs;
- planning guidance on EMF issues is thought necessary;
- undergrounding electricity lines is known to be costly;
- LPAs consider themselves to be in a weak position with regard to the planning of HVDT installations;
- it is acknowledged that new OHLs may be required to connect new gas-fired power stations.

9.6 CONCLUSION

This chapter has described a cross-case analysis based upon the results of the individual case studies. This analysis has, firstly, synthesised the components shaping the policy in the different case studies into a more generalised picture of policy formation, thus presenting the influential features, role of actors, elements of policy content and inter-linkages which are most dominant throughout the body of case studies. Secondly, the analysis has drawn out of this generalised picture three broad policy themes, which are then taken as a basis for further description of the influences at work in policy formation. Strong patterns of policy formation were revealed within the different themes, brought into sharper focus by a comparative study of the components at work within them.

Table 9.6 LPAs with no HVDT-related policy concern

LPA	Features	Planning Officer Remarks
1. Reading Borough Council	*Cf. Swindon Southern England, expanding town; OHLs present.*	OHLS not an issue.
2. Sheffield City Council	*Cf. Rotherham South Yorkshire; OHLs present, regeneration efforts.*	OHLs have been an issue. Concern arose in the mid 1990s because of housing being built underneath an OHL; anecdotal evidence suggested related ill health. An officer concern –reading on the subject. A policy was seriously considered, but not pursued on learning the electricity industry would oppose it, after a meeting with Yorkshire Electricity and Rotherham Metropolitan Borough Council; we didn't have the resources for an expert witness. Also, no other housing sites were planned near OHLs.
3. Enfield London Borough	*Cf. Newham London; Lea Valley; OHLs present.*	OHLs along the Lea Valley: industrial and recreational area; road constructed underneath OHL along one length. Proposed National Athletics Stadium – OHLs not raised as an issue. A major housing development has an OHL through the middle; corridor left as parkland. Other problems with the site, including radioactivity, were more significant. No suggestion of making OHLs a policy issue. Mobile phone masts becoming a major concern.
4. Derwentside District Council	*Cf. Tynedale North-east England; rural; designated areas; OHL present.*	Not a policy concern. OHLs pass through some conservation sites; have always been there.
5. Kent County Council	*Cf. Norfolk Eastern England; energy generation; OHLs present.*	Many OHLs along the Thames Estuary because of history of power stations; used to them. May become an issue with further energy developments. Sometimes consulted by the electricity industry, but we have very little power; e.g. cost of undergrounding gets thrown back at the local authority.
6. Yorkshire Dales National Park Authority	*Cf. Snowdonia National Park.*	No policy concern.

LPA	Features	Planning Officer Remarks
7. York City Council	*Cf. Redcar & Cleveland* *North Yorkshire; controversial OHL project.*	We feel that General Permitted Development Order (GPDO) and Electricity Act cover the issues. If an application came in, would look at visual impacts and health fears as material considerations.
8. Cotswold District Council	*Cf. Mendip* *South-west England; rural; designated areas; OHL present.*	Health concerns. Lack of planning guidance. Issue treated on a site-by-site basis. Housing development bounded by a 400 kV OHL; delayed, awaiting guidance that never arrived. We established a cordon sanitaire relating to the swing distance of the line. Possible area in the review of the plan.
9. East Hertfordshire District Council	*Cf. Aylesbury Vale* *South-east England; development pressure; OHLs present.*	Not a problem; OHLs not intrusive.
10. Newport County Borough Council	*Cf. Bridgend* *South Wales; small town, OHL present.*	Policy considered; but Swansea's policy was challenged and ruled against. Housing site affected by an OHL; 40-m safeguarding zone in site considerations – not yet challenged. Health and amenity concerns. Nothing conclusively proven; like telecommunication masts. Visual considerations on Newport Levels and Sites of Special Scientific Interest (SSSIs). Large number of OHLs due to power station, steelworks and other industry. Have been there for a long time. Aware of the cost of diverting or undergrounding OHLs. An issue we've avoided, but have discussed.
11. Braintree District Council	*Essex: relatively low level of policy concern.*	HVDT not a matter of concern.
12. Chichester District Council	*South coast: relatively low level of policy concern.*	Not considered a policy matter. Only raised with regard to conservation areas, not high-voltage lines.
13. Mid Devon District Council	*Mid Devon: relatively low level of policy concern.*	Not a pressing issue. There are existing National Grid lines, but no likelihood of new ones. A general policy on minimising the impact of development could be used with regard to new lines. Being advised to keep local plans (LPs) shorter and more focused. Though members did introduce a clause on telecommunications.
14. Newark & Sherwood District Council	*East Midlands: relatively low level of policy concern.*	OHLs have never been an issue. A power station in the District and others along the Trent Valley; been there a long time – part of the local economy.
15. North Shropshire District Council	*West Midlands: relatively low level of policy concern.*	No specific concern.
16. Salford City Council	*Manchester: relatively low level of policy concern.*	There have never been problems regarding OHLs and nearby development. Not raised as an issue by members, officers or countryside wardens.

LPA	Features	Planning Officer Remarks
17. Sandwell Metropolitan Borough Council	*Birmingham: relatively low level of policy concern.*	Rely on the planning application process. Not relevant in the Unitary development plan (UDP).
18. South Holland District Council	*Fens: relatively low level of policy concern.*	OHLs becoming relevant with the construction of a new gas-fired power station in the District and possibly others following the lifting of the moratorium; renewable projects are also likely. This has led to a landscape character assessment of the District. Ambivalence towards the landscape – people haven't appreciated it, there's no sense of ownership, due to agricultural intensification. But people are starting to take notice of the landscape.
19. South Tyneside Metropolitan Borough Council	*Tyneside: relatively low level of policy concern.*	One stretch of OHL has been removed and a substation decommissioned. Policy on telecommunications, but not OHLs.
20. West Dorset District Council	*Dorset: relatively low level of policy concern.*	No attention given; used to the OHL. More concern over mobile phone masts and radio antennae in the AONB.

Consistently occurring determinants of policy formation have therefore been brought to light by this cross-case analysis, some of which are specific to individual policy themes, others of which cut across them. The following chapter investigates these elements in more depth, drawing upon not just the results of this case study analysis, but also of the analysis of DPPs across England and Wales presented in Chapter 5. Reference is also made to those authorities that have not demonstrated HVDT-related policy concern.

CHAPTER 10

The Formation of HVDT-related Policy

10.1 INTRODUCTION

The empirical work presented in the previous five chapters has been based on the conceptual framework shown in Figure 1.2. According to this model, features that are external to the policy-making process bear on various actors, who then formulate policy concerns. These concerns give shape to more specific policy content, although the interplay of actors and concerns throughout the process of development plan preparation continues to shape the statement of policy.

This understanding has been applied to the study of HVDT-related policy at two levels: across England and Wales as a whole, and within localised settings. This has indicated a number of consistently occurring characteristics, especially in policy content, as well as revealed different aspects of policy formation at the two levels of study. This chapter draws together and discusses further the results presented in the preceding chapters, in order to construct a fuller overall picture of policy formation. This involves consideration of the analysis of development plan processes (DPP)s across England and Wales (Chapter 5) and the case study material. In relation to the case study results, the individual cases (Chapters 6–8), as well as the cross-case analysis (Chapter 9) are referred to. This is done in order to regain the detail and interest of individual cases that tended to be lost in the cross-case analysis. Reference is also made, throughout this chapter, to broader policy frameworks and other literature that relate to the points being considered.

This chapter is organised along the lines of the three policy themes described in Chapter 9. As has been shown, these relate to distinctive concerns and provide a key to the understanding of HVDT-related policy; the drawing together of results, as outlined above, is therefore conducted in relation to each of the themes. The discussion of each theme is shaped by the components shown in Table 9.5 and Figures 9.1, 9.2 and 9.3.

COUNTRYSIDE PROTECTION

Table 9.5 and Figure 9.1 show the components of the "countryside protection" theme; these form the basis for the following headings. Firstly, rural and designated areas considered sensitive to high-voltage distribution and transmission (HVDT) installations are looked at, as is the effect of installations on the landscape. Particular attention is given to the possible consequences of current HVDT developments. The environmental focus of this policy theme is then explored, especially its association with environmental strategies in plans, the role of policy makers and the influence of environmental bodies. Aspects of policy content are then examined, both the effects of installations in the countryside and the measures proposed to offset those effects. Finally, precedents and policy frameworks that are used in support of policy are considered. The intervention of certain bodies during plan preparation and the response of local planning authoritys (LPA)s to these representations are referred to throughout the discussion.

10.2 RURAL AND DESIGNATED AREAS

Within the "countryside protection" theme, the presence of extensive designated and rural areas is a universal feature (Table 9.5). This is exemplified by the case study authorities that raise this theme, all of which have

- extensive rural tracts: these cover a wide range of countryside types, including cultivated lowland, hilly pastureland, moorland, forest and mountains;
- significant sites and areas designated for a wide variety of purposes; these include those designated locally, for landscape, conservation, and other purposes: for example, Special Landscape Area, County Wildlife Site, and those designated nationally, for landscape, scientific, conservation, planning, and other purposes: for example, area of outstanding national beauty (AONB), Heritage Coast, SSSI, National Nature Reserve and Green Belt.

The latter designations include some of international importance, for example, Ramsar Site and World Heritage Site.

The "countryside protection" theme is explicit in seeking to preserve these areas, especially those that are designated, from HVDT installations; they are seen as potentially under threat and sensitive to the possible incursion of installations. Most attention is upon the protection of areas that have received national designation. However, some authorities also seek to protect their own locally designated areas; for example, Tynedale makes no distinction in its deposit plan between the level of protection sought for the North Pennines AONB and the locally defined Areas of High Landscape Value (Section 7.3). Some reference is also made to non-designated rural areas; so Aylesbury Vale states its opposition to overhead lines (OHL)s "in the countryside" (Figure 8.9).

Within the case studies, the electricity industry did not bring any objections against policies that sought only to protect nationally designated areas. This reflects the industry's own approach of seeking to minimise effects on statutorily designated areas. For instance, National Grid's (NG)'s Schedule 9 statement states that, in planning any new transmission equipment, NG will seek to avoid sites carrying major designations (Section 2.10.1) (although this does not apply to green belts, which aim to limit urban expansion rather than preserve landscape, etc.).

There is, therefore, a degree of consensus between the electricity industry and LPAs with regard to the protection of areas designated nationally for landscape , and other purposes. However, this does not dissuade some LPAs from advancing policy on this topic. This suggests that LPAs are unaware or unsure of the industry's own current practice in this regard; this is possibly due to the historic incursion of HVDT installations into nationally designated areas, such as Snowdonia (Section 7.12.1). Alternatively, this may be a statement of the high priority that LPAs give to their designated areas, even if the likelihood of a threat is remote. This would be in keeping with the sustainability strategies to which the presence of designated areas is strongly connected (Figure 9.1; and Section 10.6.1).

In contrast, the electricity industry did object to proposed policies that widened their scope to locally designated areas. This is illustrated by Redcar & Cleveland, where NG and the distribution network operator argued that the areas where the council sought to resist OHLs were so extensive that they amounted to a "blanket opposition" to OHLs in the borough (Section 8.3). This demonstrates the industry's feeling that having to avoid areas in the countryside other than those with national designation would be unduly restrictive and would hamper the provision of an efficient electricity network. In other words, in locally designated areas (and in non-designated countryside), the balance of statutory responsibilities weighs more on the side of providing an efficient electricity network than on the side of preserving amenity (Section 2.3.1). The industry has found broad support from inspectors on this point; for example, the inspector of the inquiry held into Redcar & Cleveland's plan recommended that opposition to OHLs in locally designated areas should be made less restrictive (Section 8.3).

Throughout England and Wales as a whole, "designated areas" is the policy concern that has attracted the smallest proportion of objections from NG (Figure 5.5). This relatively low level of contestation between LPAs and the electricity industry corresponds with the case study picture presented above: the policy focus is more upon nationally designated areas and is not contested; only references to locally designated areas raise objections.

On the other hand, some positive support for the protection of designated areas was expressed by environmental bodies (see Section 10.5.3). They did not suggest that protection should only extend to nationally designated areas; in fact, in the case of Mendip, the Countryside Agency reckoned that policy should be extended to cover not only locally designated areas but also "countryside and settlement character" (Section 8.7).

Throughout England and Wales as a whole, 84 DPPs show concern for designated areas (Figure 5.2); local plan (LP) and structure plan (SP) processes figure more highly in this respect due to their more rural focus (Figure 5.4). The distribution of these DPPs is presented in Figure 5.8. This shows that this policy concern is strongest in some parts of England and Wales where designated areas are numerous and extensive, such as the north Pennines and East Anglia, but not in other locations where designated areas are equally important, such as the Cotswolds and the Chilterns. The map also shows that the highest levels of contestation with NG are found in parts of northern and eastern England.

There is, therefore, significant concern across England and Wales to protect designated areas from HVDT installations. However, the presence of designated areas in itself is not sufficient to provoke policy formation; there are clusters of concern that cannot be linked simply to the concentration of designated areas. Within these clusters are "hotspots" where LPAs are attempting to defend areas designated not just nationally, but also locally. It is likely, therefore, that other factors contribute to this policy concern, such as a sense of threat to designated areas from future HVDT installations (see Section 10.4).

The protection of locally designated areas and non-designated countryside from the effects of HVDT installations remains a topic of disagreement between the electricity industry, LPAs and some agencies, and one that could benefit from greater dialogue between parties. There is, for instance, potential in investigating the possibilities for a more graded approach to the protection of rural and designated areas, as implied by inspectors, and also for determining the ability of different countryside character areas to accommodate HVDT installations, as suggested by the Countryside Agency.

10.3 ESTABLISHED HVDT INSTALLATIONS IN THE LANDSCAPE

The "countryside protection" theme is also characterised by prominent HVDT installations already affecting countryside and designated areas (Table 9.5). These include OHLs that traverse rural areas for considerable distances. Many of these installations have been in place for some time, forming part of established transmission or distribution networks. As such, they are not always considered to be intrusive in the landscape, as indicated by a Bridgend officer: "I would say that most of them [OHLs] have been there for so long, people are used to seeing them and they don't tend to notice them" (Section 8.16.1). However, the landscape setting in question is one that the officer considers capable of absorbing OHLs, saying of the line illustrated in Figure 8.13, "the fact that it dives in and out of valleys … it's not that noticeable" (Section 8.16.1).

This is in stark contrast to Snowdonia, where strong public opposition to energy infrastructure, including major OHLs, has persisted over several decades. Here, OHLs are perceived as prominent and alien features on the landscape, partly because of their dominance on certain skylines (Section 7.12.3; Figure 7.9). In a similar manner, the Bridgend officer placed priority on protecting the south Wales coalfield plateau from the far-reaching visual effects of features such as OHLs (Section 8.16.1). The degree to which existing HVDT installations in rural areas are likely to motivate policy depends, therefore, on their particular impact on the local landscape, which relates in turn to the ability of that landscape to accommodate those installations.

However, the way in which the effects of HVDT installations on the landscape are perceived appears to depend not just on the landscape's intrinsic character, but also on the value that is attributed to local landscape. For example, the insistence that Snowdonia should not suffer the visual intrusion of major OHLs has been repeatedly based on its national park status: "We're a national park for a purpose, and I think one of the purposes is to try to preserve little pieces of wilderness, or semi-wilderness where we can" (Section 7.12.3). A high regard for local landscape was also demonstrated in Norfolk, though this was stated in terms of cultural importance rather than formal designation: "the long views over open countryside are something which are particularly valued locally" (Section 7.8.2).

The localised nature of landscape appreciation can be illustrated with reference to Figure 5.7. This shows that district councils in most of Norfolk, and elsewhere in East Anglia, express policy concern for "visual effects", whereas a large block of authorities to their northwest show no such policy concern. This is despite the fact that they cover low-lying Fenland, where OHLs might be considered just as, if not more, intrusive as in the rest of East Anglia. The lack of policy expression may be due to different attitudes towards landscape; for example, an officer from a Fenland authority with no HVDT-related policy remarks that there is an ambivalence towards the local landscape: "people haven't appreciated it, there's no sense of ownership" (Table 9.6).

These examples illustrate again the relativity of the visual impact of HVDT installations to landscape type and confirm the need for a detailed assessment of the effects of installations in different landscapes. However, they also suggest that any assessment should take into account not just the physical components of landscape, as in the Countryside Agency's "character area" approach (Countryside Agency 1999), but also public perceptions of landscape; an understanding of landscape is required that incorporates the values attributed to different areas.

10.4 CURRENT HVDT DEVELOPMENTS: HEIGHTENED CONCERNS

Within the "countryside protection" theme, there is a frequent occurrence of new HVDT installations being given active consideration (Table 9.5). This is either because installations are recently (or currently) being developed or because they are envisaged as possible future constructions.

Recent or current projects involve either new lines connecting new generating plant to grid systems or the upgrading of existing equipment. A number of these projects have given rise to local opposition; this has been expressed most strongly through interest groups, though some wider public opposition and member concern is also evident. For example, a relatively small-scale project in Tynedale provoked controversy; here, a proposed substation was resisted by residents and amenity groups on the grounds of its visual intrusion and impact on designated areas (Section 7.4.1). Although this case did not play a crucial role in policy formation in the district, it reinforced existing concerns about the possible effects of HVDT installations.

10.4.1 The "dash for gas"

More substantial projects have been carried out during the last few years in relation to new gas-fired power stations. Both Redcar & Cleveland and Norfolk reflect this trend; major OHLs are under construction in both of these areas as a direct result of new or upgraded power stations exploiting North Sea gas (Sections 8.4.1; 7.8.1). These power stations are symptomatic of the "dash for gas" that took place in the 1990s, as the consequence of a number of factors (Thomas 1996, Patterson 1999):

- liberalisation of the electricity supply industry, permitted by privatisation of the industry in 1990;
- newly developed combined-cycle gas turbine plants, with major technical and environmental advantages over coal-fired power stations;
- a political desire for reducing reliance on coal as an energy source.

The electricity transmission network has therefore had to adapt to a major shift in the pattern of energy generation over the last decade, by providing links to the new power stations. Given the fairly comprehensive coverage of the network across England and Wales, new lines connecting gas-fired plant to the grid have been of a relatively short distance (with the exception of the "North Yorkshire" line (Section 8.4.1)). However, they have involved a significant number of LPAs in consent procedures for these projects and have increased sensitivity to the possibility of future applications for major OHLs.

By 1998, the share of total electricity generation provided by gas-fired plant had increased to about 30% and was projected to increase to over 50% by 2003 (Department of Trade and Industry (DTI) 1998). Misgivings about over-reliance on gas and further decline of the coal industry were expressed by the newly formed labour government; this led to the introduction of a stricter consents policy on new gas-fired power stations (DTI 1998). This was not, however, a moratorium on gas-fired power stations, and a considerable number of proposals remained in the pipeline at the time of this policy review. Moreover, this approach has since been relaxed, and dependence on imported gas supplies is likely to increase in coming decades (Performance and Innovation Unit 2002). The construction of combined-cycle gas turbine plants, along with associated connections to the transmission network, is therefore continuing apace in parts of England and Wales.

Early experience of LPAs during this period is illustrated by Cleveland County Council. The council initially welcomed the proposal for a major gas-fired power station on Teesside, but later felt "caught out" when the unforeseen applications for connecting transmission lines were submitted (Section 8.4.1). In contrast, Norfolk, reviewing its SP in the late 1990s, made specific reference to electricity lines that would be required in association with new power stations (Section 7.7).

This suggests that the decade witnessed an increasing awareness by LPAs of the consequences of consent being granted for major new power stations; this is particularly at county level, where a more strategic approach to energy issues is to be expected. More generally among LPAs, there has been a build up both in the local experience of transmission lines required for the new gas-fired power stations and in the perception that such lines are a future possibility. This has had a direct bearing on HVDT-related policy in certain areas, heightening concern for the protection of countryside and designated areas; this is sharply illustrated by Redcar & Cleveland and its neighbouring authorities (Section 8.4.1; Figure 5.8).

10.4.2 Renewable energy schemes

A second sector in which HVDT installations are receiving active consideration is that of renewable energy schemes. This is the case in three of the case study authorities:

- Norfolk has recently seen wind turbine and biomass-fuelled projects developed in its area, and it expects further proposals (Section 7.8.1);
- Snowdonia is aware of pressure for the exploitation of wind, water and biomass resources in and around the park (Section 7.12.5); moreover, it has a long-standing history of the use of water power (Section 7.12.1);
- Mendip perceives wind turbine projects to be a possibility in its area (Section 8.8.1).

Renewable energy schemes are generally of a much smaller scale than conventional power stations and have lower-voltage outputs. As a consequence, associated electricity lines are smaller, sometimes underground, and usually connect directly to distribution networks rather than to the transmission grid (NG 2001c). However, these schemes also tend to be more dispersed and are often located in areas of considerable landscape value (DoE 1993b). LPAs have therefore expressed concern for the visual and other effects of renewable energy projects and their connecting lines, especially in rural and designated areas. This is illustrated by Mendip, where the notion of wind turbines being developed has motivated policy on protecting designated areas from OHLs (Section 8.8.1).

The development of renewable energy schemes is taking place against a strong national policy background. Government targets for the amount of electricity to be derived from renewable sources have been in place since 1990 (DoE 1990b) and is periodically increased; a recent target of 10% of total electricity generation by 2010 (DTI 2001) has already been surpassed by a more ambitious target of 20% by 2020 (Performance and Innovation Unit 2002, DTI 2003). Mechanisms for achieving these targets have also been advanced, requiring set quotas of electricity to be obtained from renewable sources (initially the Non-Fossil Fuel Obligation, now replaced by the Renewables Obligation). National planning guidance advises local authorities to consider the contribution they can make to renewable energy targets (DoE 1993b), and regional studies have been carried out to carried out to assess the potential of different forms of renewable energy in each region (e.g. Energy Technology Support Unit 1997).

LPAs are showing an increasing awareness of and responsiveness to this policy drive. This is illustrated by Norfolk and Snowdonia, which make specific reference in their plans to national renewables targets and respective regional studies (Sections 7.8.1; 7.12.5). Snowdonia's description of Non-Fossil Fuel Obligation contracts as creating a "rush to capitalise on the renewable premium" (Section 7.12.5) is indicative of a sense of entrepreneurial, and perhaps questionable, activity within the sector. It is, therefore, in this context of this upsurge of interest and action in the renewables sector that some LPAs are establishing policy on the siting of potential schemes, and are including the possible effects of associated lines as a consideration to be taken into account. National planning guidance gives them direct encouragement to do this, as it advises LPAs to consider in their development plans the environmental effects of renewable energy projects, especially upon designated areas and in green belts (DoE 1993b, Welsh Office 1999). The growth of the sustainability agenda within plan-making has also encouraged LPAs in this respect; in particular, the inclusion of "biospheric" issues, such as global warming, in plans has encouraged LPAs to give greater consideration to renewable energy (Section 3.6.4).

This trend is likely to be strongest in areas where there is known to be the greatest potential for renewable energy schemes – as revealed by the regional studies, for example. There is also evidence, however, that broader public awareness of the likely increase in renewable energy supply, especially from wind turbines, in itself gives rise to concerns about the possible associated development of OHLs; this was clearly the case in Mendip (Section 8.8.3). This is in line with the growing public resistance to renewable energy projects, which itself influences the planning system, as has been noted by the Royal Commission on Environmental Pollution (RCEP) (2000, p. 149).

10.4.3 Geographical distribution of policy concerns

The influence of these two recent developments in electricity generation upon HVDT-related policy is also suggested by the geographical distribution of policy concerns. Figures 5.7–5.9 reflect the "countryside protection" theme most clearly, which has been highlighted above as the area of concern most motivated by the development of gas-fired power stations and renewable energy schemes. These maps show large clusters of DPPs in parts of England and Wales where these forms of electricity generation are becoming established:

1. the east coast of England, reaching inland from Teesside, Humberside, the East Anglian coast and the Thames Estuary: these are all areas where gas-fired power stations have recently been constructed, taking advantage of North Sea gas pipeline terminals;
2. throughout northwest and southwest England, and Wales: these are all areas where wind energy projects have recently been built and where potential for

further schemes is greatest, owing to the presence of upland areas exposed to prevailing westerly winds.

There is, therefore, a broad correspondence between these policy distributions and the patterns being established by new forms of electricity generation. This suggests that current trends in energy policy have a direct bearing on the formation of certain aspects of HVDT-related policy in England and Wales.

10.5 ENVIRONMENTAL PRIORITIES

10.5.1 Environmental strategies

Within the case study authorities, the "countryside protection" theme has strong links to strategies adopted for plans as a whole, more so than the "residential area protection" theme (Table 9.5). In all cases, the theme is associated with strategies of sustainability or environmental protection (Table 9.2). For example, Tynedale sets out an environmental strategy for its plan, including the need "to conserve the District's natural and historic assets"; its policy on OHLs clearly reflects these priorities (Section 7.4.2). Likewise, Bridgend sees itself as having particular strengths with regard to sustainability, and mentions its concerns regarding HVDT installations within the context of environment protection (Section 8.16.1).

These broad strategies are themselves an expression of an environmental/sustainability agenda that has become increasingly central to development plans since 1990 (Section 3.6.3). The "countryside protection" theme results, in part, from this trend; it is an expression of the increased importance being attached to preserving the features of designated areas and the wider countryside. This is especially so, given the landscape emphasis of the policy theme, as this relates closely to those "traditional" concerns of planning that were taken up most easily and quickly into the new environmental agenda (Section 3.6.4). Several of the case study authorities have, in fact, carried out formal landscape assessment of their areas, in order to ensure greater environmental protection (Figure 9.1); at Redcar & Cleveland, for example, it is intended that development should respect local landscape character (Section 8.4.2). The "countryside protection" theme is closely tied to this growing emphasis on environmental quality, especially to the conservation aspects of the new agenda.

In some cases, specific reasons are given for seeking to preserve countryside and designated areas, other than their implicit qualities (Table 9.2). These include tourism (Section 7.4.2), regeneration efforts (Section 8.4.2) and the preservation of mineral reserves (Section 8.16.2). Similarly, external pressures on the countryside give extra force to environmental strategies, such as development pressure (Section 8.12.1). There are, therefore, a range of economic and social aims that LPAs combine with their environmental strategies, which provide added underlying motivation for HVDT-related policy.

10.5.2 The role of planning officers and elected members

The development of this policy theme has been largely in the hands of planning officers. Compared to the "residential area protection" theme, there has been relatively little instigation of policy from elected members or active concern expressed by the public at large (Table 9.5; Section 10.9.1). Policy initiatives have generally originated with officers drawing on their own sense of professional opinion. This is indicated by the clear, and sometimes strong, views on the landscape effects of OHLs expressed by officers in a number of the case study authorities. A Tynedale officer, for example, referred to OHLs as "intrusive elements" whose "pervasive nature reduces landscape quality" (Section 7.4.2); the views of other officers are also referred to above (Section 10.3). There is, therefore, a sense of expertise at officer level with regard to the conservation of landscape within their areas, which is possibly linked to the formal landscape assessments mentioned above. It is also likely to be a reflection of the emphases in the training of planning officers. Officers therefore demonstrate a confidence in setting out policy covering the "amenity" aspects of HVDT installations (Section 2.3.3). There is also evidence that in-house landscape staff have been involved in HVDT planning issues (such as in Norfolk (Section 7.8.2)), adding to the sense of capability in setting out criteria for the location of HVDT installations.

Although not a common element in the case study authorities, some elected members have also demonstrated commitment to the "countryside protection" theme. Members at Redcar & Cleveland and Norfolk showed particular concern for landscape quality (Sections 8.4.2; 7.8.2). Norfolk members went against the advice of officers in opposing proposals for a new OHL on visual grounds (in a manner similar to member insistence in pursuing health concerns (Section 10.9.3)). There is in certain areas, therefore, a high degree of landscape appreciation at member level, probably reflecting local cultural attitudes towards landscape (Section 10.3), rather than the professional approach of officers.

Additional features in Snowdonia were the backing for policy provided by national park bodies, and the formation, in 1974, of a park authority with full planning powers (Table 9.2; Section 7.12.2). This highlights the importance to policy formation of institutional arrangements that are capable of representing a locality's interests.

10.5.3 Environmental bodies and groups

This policy theme is the only one to have received comment from national environmental bodies (Table 9.5). This has been in the form of mostly supportive representations at consultation stages of plan preparation, from the Environment Agency (Section 8.11), the Countryside Agency (Section 8.7), English Heritage and the Royal Society for the Protection of Birds (Section 7.3). Their comments have been of a brief and general nature and do not demonstrate a very significant concern with the effects of HVDT installations.

Greater attention has been paid by environmental and other interest groups to proposed projects in several of the case study areas, as referred to above (Sections 10.3; 10.4). Major contentious projects have aroused organised opposition, in the past, in Snowdonia (Section 7.12.2) and, recently, in North Yorkshire and Teesside (Section 8.4.1). Existing bodies have become involved, such as the Council for the Protection of Rural Wales and newly formed action groups, such as REVOLT. Although the focus of their attention has been to campaign against specific proposals rather than to try and influence policy, they have raised wider awareness in their localities about the possible effects of HVDT installations, which has then fed through to policy formation. This has been most clearly the case in Redcar & Cleveland, where effective publicity has succeeded in making OHLs into "quite an emotive issue locally" (Section 8.4.1). Interest groups have, in fact, exerted their greatest, albeit indirect, influence on HVDT-related policy in relation to this policy theme (Table 9.5).

Environmental groups have therefore tended to react to particular instances of OHLs affecting the countryside rather than seek to offset future intrusion through involvement in plan-making. Their influence on HVDT-related policy has depended upon planning officer receptiveness towards, and willingness to represent, their concerns; this may rely, to some extent, on a mutual reinforcement of interest. At Mendip, for example, a sense that the local Council for the Protection of Rural England (CPRE) group and others were concerned about the visual impact of OHLs explicitly encouraged officers to develop policy on the issue (Section 8.8.1) – in line with the council's emphasis on public participation in plan-making (Section 8.8.3). This suggests that environmental groups stand to benefit from taking a more active role in the plan-making process and possibly in seeking other formal settings in which to express their concerns (as occurred in relation to "residential area protection" (Section 10.9.1)).

At a national level, both the CPRE and the Countryside Commission (now the Countryside Agency) have in recent years tried pressing for measures that would reduce the visual effects of OHLs in the countryside, especially the undergrounding of lines (CPRE 1995, Countryside Commission 1994). However, these bodies are not currently pursuing this issue, possibly due to a lack of success in bringing about any change in the industry's practice (see Section 10.6.2). REVOLT has now taken on a leading role in campaigning against the visual and perceived health effects of OHLs, having widened its scope beyond the North Yorkshire project that it was initially formed to oppose. Action is also being led by the Friends of the Lake District, which has commissioned a report on the possibilities for undergrounding lower-voltage lines. It is likely that activism of this kind will continue to arouse concerns about the effects of HVDT installations in the countryside and subsequently feed through to localised policy formation, as has happened over the years in Snowdonia (Section 7.12.2).

10.6 POLICY CONTENT

10.6.1 Effects of HVDT installations in the countryside

As discussed in Section 10.2, the "countryside protection" theme is concerned with the effects of HVDT installations in areas of high landscape, conservation, and other values, especially in areas designated for such reasons. The particular effects in mind are now considered.

With regard to landscape, the policy theme concentrates primarily on the visual quality of certain areas (Table 9.5); installations are described, for example, as having a "dominant visual impact on the landscape" (Figure 8.12). This emphasis is evident across England and Wales, where "visual effects" is the most frequent policy concern (Figure 5.2). At first sight, this implies a relatively limited notion of landscape, one that takes into account the views across an area, but not other aspects of a place, such as cultural, wildlife, seasonal and other elements (Landscape Institute and Institute of Environmental Management and Assessment 2002). However, a broader understanding of landscape is implied in the listing of sites designated for a wide range of purposes, including ecological, archaeological, historic and other interests (Table 9.5). Implicitly, if not always explicitly, LPAs are seeking to protect the overall qualities of places from HVDT installations; hence Snowdonia states its aim of "conserving the natural beauty of the Park" (Figure 7.8).

Where ecological, archaeological, and so on, sites are referred to, no mention is made of any specific effects of HVDT installations, such as the danger of birds colliding into OHLs (Table 2.1). This suggests that no detailed analysis has taken place of the ways in which HVDT installations might impact upon the countryside and designated areas. This is further illustrated by the fact that the possibility of damaging archaeological remains by placing lines underground (the most frequently proposed policy measure) has not always been appreciated by LPAs (such as at Tynedale (Section 7.3)). It appears that opposition to HVDT installations in the countryside has been based upon general impressions of harm rather than a consideration of specific impacts; LPAs may well benefit, therefore, from the more careful analysis of possible effects undertaken by the electricity industry, such as that presented in environmental reports (Section 2.8).

There are also passing references to other possible effects of HVDT installations in the countryside (Table 9.5). Firstly, Norfolk makes reference to potential effects upon recreation, tourism and amenity (Figure 7.4); secondly, Bridgend seeks to prevent OHLs from sterilising any mineral reserves in its area (Section 8.16.2). These considerations, especially the second, touch upon economic attributes of the countryside, rather than the aesthetic and conservation values underlying the other effects described above. Nonetheless, they are treated by the LPAs concerned along with the other issues covered by this policy theme.

The electricity industry has generally objected neither to the implication that HVDT installations may have harmful effects in the areas stipulated nor, in the case of nationally designated areas, to the principle of seeking to protect them (Section 10.2); in many ways, the concerns of LPAs echo those of the Amenity Clause (Section 2.3.3). Objections have focused rather on the measures of protection that have been proposed.

10.6.2 Undergrounding

Undergrounding is the most common policy measure proposed in the "countryside protection" theme (Table 9.5). Typically, policies call for new lines to be placed underground as cables, rather than suspended overhead from towers, especially when crossing designated areas. So Mendip states, "Where there is no alternative to power lines crossing sensitive areas, they should be placed underground ..." (Figure 8.5). There is no suggestion that existing OHLs should be placed underground, and there is recognition that undergrounding may not always be feasible in the specified areas. But there is, in some cases, a determined tone to policy; for example, Snowdonia states that it "will insist that Supergrid power lines are laid underground where they need to pass through the Park" (Figure 7.8).

Throughout England and Wales as a whole, 111 DPPs show concern for "undergrounding", the second most frequent policy concern (Figure 5.2). The distribution of these DPPs is presented in Figures 5.6 and 5.9. This shows a very similar pattern of policy concern to "designated areas" (Figure 5.8), though more widespread. This reflects the application of undergrounding to other concerns as well as the protection of designated areas; for example, it is also proposed as a solution in the "encouraging regeneration" theme (Section 10.12.2). LPAs have, therefore, a widespread preference for placing lines underground. This might seem surprising, given the lack of statutory control they have over the laying of cables (Section 2.4.2), but indicates their strong perception that undergrounding lines can offset the effects of OHLs.

<u>Objections to calls for undergrounding</u>

However, this policy measure has been the focus of a relatively high number of objections from NG (Figure 5.5). For example, NG objected to Bridgend's proposed policy, stating that the undergrounding of lines presents serious financial, technical, ecological and archaeological difficulties (Section 8.15). NG has gone to some length to present these arguments more widely, through the publication of a booklet comparing the relative strengths and weaknesses of OHLs and underground cables (Section 2.10.2). This points out, for example, that the cost of high-voltage cables can be 20 times greater than OHLs and that undergrounding can cause greater environmental damage through disturbance to land; there are also pollution issues associated with underground cables (Table 2.1). NG's policy is therefore to underground lines only in exceptional circumstances (NG unpublished.d). Some of the case study authorities accepted these arguments and modified their policies accordingly, such as Bridgend (Figure 8.12). In fact, a relatively small proportion of

LPAs throughout England and Wales attempted to defend their calls for undergrounding at inquiry (or examination) stage (Figure 5.5), suggesting that most were amenable to the arguments put forward by NG.

NG's reservations about undergrounding are largely supported by wider attitudes within the electricity industry; International conference on large high-voltage electric systems' (CIGRE's) comparative assessment of underground cables and OHLs concurs with that of NG (Table 2.4). However, there are international examples of greater use being made of undergrounding, such as in Denmark, relying partly on technical advances (Section 2.11). This suggests that there may be potential in the future for the wider application of undergrounding and that as well as highlighting the practical difficulties involved, the electricity industry could remain open to the possibility of means of overcoming them.

10.6.3 Other proposed measures

The other main measure that is suggested to protect countryside and designated areas is opposition to any new HVDT installations passing through the stipulated areas. As discussed above (Section 10.2), this met with objections from the electricity industry when extended to areas other than those statutorily designated.

The implication of this approach is that lines should be sited away from the stated areas. This inference is made explicit by Snowdonia, which insists that "new lines are routed around the Park" (Figure 7.8). This is stated, however, without any consideration for the implications for surrounding areas. This was an issue that a Redcar & Cleveland officer commented upon, in relation to the adjoining North York Moors National Park: "We're very concerned that we don't get the overspill from the Park" (Section 8.4.2). This highlights the localised nature of this policy area, whereby an LPA may feel obliged to defend its territory in order to match the protective stance of its neighbours. It also suggests that a strategic approach to the planning of HVDT installations is necessary to overcome the uneven assertion of local interests, particularly with regard to lines potentially affecting more than one LPA. It could be argued that such an approach already exists in relation to permission for OHLs, as this is currently granted at government rather than local level (under "Section 37 consent" (Section 2.3.2)). However, this does not give LPAs the opportunity to participate in strategic discussions about route selection either with one another or with the electricity industry; it may be that there is potential for a more collaborative approach to the identification of route corridors that ensures an equitable level of protection for valued areas.

Slightly less restrictive approaches are taken by some of the case study authorities, which accept that OHLs may have to cross the areas they seek to protect, but that they should be sited as sensitively as possible. For example, Aylesbury Vale states, "Where no alternative exists, the Council will seek to ensure that the lines do not intrude upon open skylines ..." (Figure 8.9). This corresponds to the electricity industry's own practice of seeking to minimise visual impacts by measures such as careful line routeing (Section 2.10.2).

A range of measures is therefore put forward in the "countryside protection" theme. They can be seen as a hierarchy of diminishing effectiveness:

- avoiding specified areas altogether;
- underground cables through these areas;
- carefully routed OHLs across these areas.

A different suggestion is made by Norfolk that new power stations should be located in proximity to the existing transmission network in order to avoid the need for extensive new OHLs (Figure 7.4; Section 10.4.1). This is a position similar to that of the Royal Commission on Environmental Pollution (RCEP), which has recently recommended that applications for new generating plant should also cover any necessary OHLs (RCEP 2000); the same idea has been put forward in the Tees Valley Structure Plan (Section 8.4.1). These suggestions represent a more strategic approach to energy installations as a whole; they also invite the question of bringing energy generation closer, not just to the existing transmission network but also to the point of electricity demand. Similarly, an Aylesbury Vale officer raised the possibility of a wider range of alternatives to new OHLs in the countryside being considered, including the questions of "whether electricity needs to be supplied or whether an existing line can be upgraded" (Section 8.12.1).

10.7 POLICY SUPPORT

10.7.1 Policy precedents

The importance to this policy theme of recent and proposed HVDT installations in LPAs' own areas has been discussed above (Section 10.4). This has given policy makers direct experience of the possible effects of installations and an awareness of future possible projects. In addition, knowledge of particular projects has provided precedents, which have then been given expression in some of the policy measures described above (Table 9.5). Notions of line routeing, for example, have specific precursors in some authorities, most markedly at Snowdonia, where a major line was routed in such a way that it passed round the park (Section 7.12.4).

The role of policy precedents has been most important, however, in relation to undergrounding. Historic cases of undergrounding took place around Snowdonia, in order to protect highly regarded landscape (Section 7.12.4); similarly, stretches of high-voltage lines have recently been placed underground in Norfolk (Section 7.8.2) and North Yorkshire (Section 8.4.1). These examples have presented the notion of undergrounding to LPAs as a technique that can be applied more widely. This measure has also been given viability in the minds of policy makers by the much more common practice of placing lower-voltage supply lines underground. This was evident at Bridgend, where the idea of undergrounding appears to have extended from smaller- to larger-scale lines (Section 8.16.1), and also at Norfolk (Section 7.8.2).

In several cases, objections from the electricity industry have led officers to a greater understanding of the difficulties involved in placing lines underground (Section 10.6.2). However, elected members have sometimes continued to press for this measure to be pursued in a manner comparable with attitudes towards EMF-related health concerns (Section 10.9.3). For instance, Norfolk members have pressed particularly strongly for new lines to be placed underground, going against the advice of their own officers (Section 7.8.1).

It would seem, therefore, that undergrounding is a policy measure that initially holds great promise for LPAs. It is a measure that they have adapted from existing practice for lower-voltage supply lines and other services, and a limited number of precedents also exist for higher-voltage lines. But it is equally a measure that they are often willing to yield, once they gain knowledge of the considerable difficulties involved in implementation.

10.7.2 Policy framework

Hierarchical policy and guidance

Several of the case study authorities have justified their stance on protecting the countryside with reference to hierarchical policy and guidance (Table 9.2). For example, Redcar & Cleveland had regard to its SP and regional planning guidance, both of which seek to protect designated areas from OHLs (Section 8.3); Snowdonia was able to appeal to national park purposes (Section 7.12.3). The level of hierarchical support is noticeably stronger within the "countryside protection" theme than in the "residential area protection" theme (Table 9.5; Section 10.9.5). This can also be seen in Figure 5.6, which shows that, across England and Wales as a whole, the policy concerns associated with "countryside protection" are much more strongly expressed at county level than those associated with "residential area protection". There is, therefore, a degree of consensus and momentum to this policy theme at different levels of planning. This is also evident at Norfolk, where the SP reinforces policies advanced by district councils (Section 7.8.2). Overall, it appears that the concerns expressed within the "countryside protection" theme are more central to planning as a whole, possibly because they relate closely to those conservationist traditions that have been taken into the new sustainability agenda (Section 3.6.4).

Major developments

One of the provisions drawn upon is that of restricting major developments in certain designated areas. Firstly, Tynedale referred to national guidance (*PPG 7*), which states that major developments should only be permitted in AONBs and national parks under exceptional circumstances, such as there being a national need for the development; the council attempted to apply the same criteria to other designated areas (Section 7.3). Secondly, Snowdonia referred to this guidance, also appealing to the "Silkin test" in this regard (Section 7.12.3). Here, NG objected to the definition of OHLs as a form of major development, but this was not pursued conclusively (Section 7.11). Whether HVDT installations would be regarded as "major developments" is likely to depend on the scale and characteristics of particular installations; the significance of this provision, however, is that it has generated policy ideas that have been taken up and extended to the consideration of HVDT installations in the countryside.

Policy history

This policy theme is also characterised by a history of policy in locally prepared plans, more so than in the "residential area protection" theme (Table 9.5). Within the case study authorities, Snowdonia has the longest history of policy on HVDT installations, beginning in the 1970s (Section 7.11), while other LPAs have precursors to their current policies in plans dating from the early 1990s (Sections 8.3; 8.11; 8.15). This demonstrates an underlying concern with the effects of HVDT installations, particularly since the inauguration of the "plan-led system" (Section 3.3.2) and suggests that, once a policy issue is established within an LPA, it is likely to be taken up for consideration in subsequent plans.

RESIDENTIAL AREA PROTECTION

Table 9.5 and Figure 9.2 show the components of the "residential area protection" theme; these form the basis for the headings given in the following text. Firstly, areas that are considered sensitive to HVDT installations are looked at and the patterns of installations that give rise to these perceptions. Secondly, given the dominance of health concerns in this policy theme, these (along with safety concerns) are examined at length; this includes consideration of the influence of the media, the role of policy makers, use made of precautionary approaches and other policy instruments, inter-authority contact and other factors. Amenity effects are then looked at, and certain aspects of proposed policy measures are explored. The intervention of the electricity industry during plan preparation and the policy response to these representations are referred to throughout the discussion.

10.8 SENSITIVE RESIDENTIAL AREAS

Within the "residential area protection" theme, the effects of HVDT installations on given areas is a dominant feature (Table 9.5) arising from the proximity of installations to residential development (Figure 9.2). As in the "countryside protection" theme, therefore, policy can be considered in terms of the defence of sensitive areas (Section 10.2). The subject of policy is partly the protection of housing, and so on, from the effects of *future* HVDT installations (so Bridgend resists the construction of new OHLs near housing; Section 8.15). But, more importantly, policy is concerned with residential areas thought to be under threat from *existing* installations. "Sensitive areas" therefore consist of areas of current (or proposed) housing already in proximity to HVDT installations. There are, in fact, two distinct patterns apparent in this regard.

10.8.1 HVDT installations in established areas

The first pattern of HVDT installations in residential areas is that of OHLs and substations serving established urban and industrial areas. For example, Rotherham and Redcar & Cleveland both have a considerable number of installations within their towns, partly to supply heavy industry (Sections 6.8.3; 8.4.1); the HVDT installations and nearby housing are well established and represent a relatively stable pattern of land use (Figures 6.7 and 8.2).

In both these cases, policy formation has been directly connected to specific residential sites. In Rotherham, for instance, policy was drafted partly in response to member concerns about the proximity of a children's play area to an OHL (Section 6.8.1). But it appears that the location of HVDT installations near housing, and so on, is not sufficient in itself to provoke policy formation; there are other authorities with a high concentration of installations in urban areas that have not pursued similar policies (Table 9.6). Clearly, there are other contributory factors, as explored below.

10.8.2 Suburban expansion around OHLs

The second pattern is that of OHLs surrounded by recent suburban development. This is exemplified by Swindon and Aylesbury Vale, where the growth of towns has gradually encroached upon OHLs located around their perimeters (Sections 6.4; 8.12.2). This is a more dynamic situation than the first pattern; recent expansion of towns has led to a new juxtaposition of OHLs and housing (Figures 6.3 and 8.10). Where this is the case, there is often further development due to take place near OHLs.

In some of the case study authorities, the juxtaposition of HVDT installations and residential areas is linked to development pressures and major housing developments (Figure 9.2). These authorities have all been given large housing allocations, reflecting regional economic growth and demands; this is symptomatic of national housing pressures, especially in southern regions (DETR 1999b). As a result, new residential estates are being constructed on the edges of many urban areas; these are precisely the locations where OHLs are most likely to be present, either feeding into the urban areas or, because they were originally sited, skirting around them. The presence of these lines has therefore become a constraint on the design of new development.

Some of the case study authorities have dealt with this issue through development control. For example, Mendip negotiated a development-free corridor along the length of an OHL bisecting an in-fill site (Section 8.8.2). A site-specific approach was also taken by Bridgend, in a development brief prepared for its major new residential area (Section 8.16.3). Particular instances of this kind appear to have contributed directly to subsequent policy formation. This is clearly the case in Aylesbury Vale, where a local example of new housing being kept at a set distance from an OHL explicitly informed the authority's development plan policy on the issue (Section 8.11). Here, the officer commented that the interrelation of OHLs and housing is "something we've always had to consider" (Section 8.12.2). The local experience of the difficulties of planning new development around OHLs therefore plays an important part in the formation of broader policy which is then seen as applicable to possible future developments.

However, there are cases where policy has been of questionable future relevance. For example, in Newham, anxieties in a neighbourhood that was bisected by a major OHL led the council to propose development exclusion zones along the length of OHLs, despite the fact that no further housing development was planned close to them (Section 6.12.1). This underlines the importance that local instances of housing near installations, and the experience of residential proximity, have in contributing to this policy theme.

10.8.3 Site-specific protection of residential areas

It is evident from both the patterns described above that this policy theme is being raised not just in response to the general possibility of HVDT installations affecting residential areas, but with specific affected areas of housing and associated land use in mind. Even where a broader approach is taken, such as by Bridgend, there are also instances of housing close to existing OHLs (Section 8.16.3). The overriding approach taken in the "residential area protection" theme is one of immediate relevance to actual or proposed residential sites. This is in some contrast to the "countryside protection" theme, where a more strategic approach is taken in seeking to protect designated areas, and so on, from the possible future incursion of HVDT installations.

The policy concerns identified across the whole of England and Wales that relate most strongly to this policy theme are "other development" and "electromagnetic field (EMF)" (Section 5.6.3). These occur most frequently in unitary development plans (UDPs), frequently in LPs, but very little in SPs, confirming their more localised and urban focus (Table 5.5). Figures 5.10 and 5.11 show their geographical distribution among LP and UDP processes. These maps show the more dispersed nature of these policy concerns and some tendency to occur in smaller, more urban areas (especially "EMF"). Although these distributions cannot be closely correlated with the patterns described above, they are consistent with them. For example, a high incidence of HVDT installations in urban areas is to be found in south Yorkshire and north Lincolnshire, where "other development" features strongly; also, development pressure is at its greatest in southeast England, where the two policy concerns are strongly represented.

There are a number of effects of HVDT installations to which residential areas are considered sensitive, according to this policy theme (Table 9.3). These fall under two headings, public health and safety, and amenity.

10.9 PUBLIC HEALTH AND SAFETY

10.9.1 EMFconcerns

Nearly all the case study authorities that raise the "residential area protection" theme express concerns about the possible effects of HVDT installations on the health of people living nearby (Table 9.5). This is typically stated in terms of public anxiety; for example, Rotherham, in its deposit plan, states: "There is growing public concern about the possible health effects of electric and magnetic fields associated with major overhead powerlines" (Figure 6.5).

Media reports

One of the most dominant features in this theme is the influence of media reports on EMF research (Table 9.5); these have contributed particularly to the raising of concerns among the public and elected members (Figure 9.2). The case studies show that, throughout the 1990s, there was a series of media reports, including press items and television documentaries, that publicised the results of research projects into the possible health effects of EMF exposure. Although these findings have not generally supported the contention that exposure to EMF emitted by OHLs can cause ill health (World Health Organisation 2000), they have provided regular opportunities for journalistic reporting. Whether or not these reports could be described as sensationalist, they have aroused strong local feelings in certain settings. In Newham, for example, residents living near an OHL organised public meetings to express their anxieties, enlisting support from their councillors (Section 6.12.1); in Swindon, feelings ran high over plans to build a school near an OHL (Section 6.4).

Public and member involvement

EMF concerns are, in fact, responsible for the highest degree of public involvement in HVDT-related policy (Table 9.5). In one or two authorities, public comment has been organised and specifically directed at plan-making; this was most notably the case in Swindon, where the local Friends of the Earth group gathered support in pressing for new development to be kept at a distance from an OHL (Section 6.4). This latter case was unusual in that an environmental forum set up by the council provided a formal mechanism for concerns to be communicated to the LPA (Section 6.4). More commonly, however, public anxieties have been expressed informally, and channelled through members, as at Redcar & Cleveland, for example (Section 8.4.3).

Similarly, members themselves have shown the greatest degree of interest in this aspect of HVDT-related policy (Table 9.5). In this regard, they have been acting partly in response to public feeling and also out of their own underlying concerns. Again, these concerns have been generated partly by media reports, such as at Rotherham: "our local members ... were picking up the press reports" (Section 6.8.1). More anecdotal accounts of the supposed harmful effects of EMF have also played a part in shaping member attitudes (Table 9.3); for example, a Bridgend member recounted the death from cancer of an acquaintance who had lived near an OHL (Section 8.16.3). In addition, members in some areas have been particularly sensitive to EMF concerns because of a wider interest in public health issues. This has been most marked in authorities with a history of heavy industry, where members have traditionally had an interest in health and safety matters; in Redcar & Cleveland, for instance, "members ... had a very good track record on things like asbestos" (Section 8.4.3).

10.9.2 Policy contestation

Electricity industryobjections

EMF concerns have proved to be the most contentious aspect of HVDT-related policy. Throughout England and Wales as a whole, "EMF" and "other development" have aroused the greatest degree of opposition from NG and the most concerted defence by LPAs (Figure 5.5). Within the case study authorities, the industry's objections to proposed policies have been based upon refuting the notion of EMF-related health risk; at Tynedale, for instance, both NG and Northern Electric objected to a proposed policy that implied that OHLs in residential and recreational areas were a health and safety risk (Section 7.3). The industry's arguments have relied upon the findings of the government-recognised National Radiological Protection Board (NRPB), that there is no proven link between EMF exposure and the incidence of cancer (Section 2.6.2), and that there is no health-based justification for restricting development in the vicinity of OHLs (Section 2.10.2).

There are, therefore, widely divergent and strongly held views on the possible implications of HVDT installations for nearby residential areas. These represent, on the one hand, a public perception of risk, reinforced by widespread media reports, and, on the other hand, technical expertise that has official backing, but which is also linked to commercial interests. It is the first of these perspectives that has motivated policy formation at the LPA level. The challenge it poses to the electricity industry is reflected by the high proportion of objections made to proposed policies; this concurs with the defensive stance taken on EMF that was noted in the review of industry-based literature (Section 2.13).

Inspector recommendations

Policies that deal with EMF concerns have therefore been the subject of disputation at a number of public local inquiries. In general, inspectors have favoured the arguments put forward by the electricity industry and have recommended the modification or deletion of proposed policies (Figure 9.2). Within the case study authorities, for example, the inspector of the inquiry into Swindon's plan recommended that health reasons should not be used as a justification for development exclusion zones around OHLs (Section 6.3). At Rotherham, the inspector recommended that the notion of a fixed-distance zone should be replaced with a more discretionary approach (Section 6.7). Elsewhere, inspectors have shown some sympathy for LPA concerns; for example, the inspector of one inquiry felt that local authority expression of concern for the health and welfare of residents was justified (Copeland Borough Council 1995). But the overriding trend has been for inspectors to place the greatest weight on NRPB's advice and recommend the deletion of references to exclusion zones (as at Calderdale, e.g. Section 6.8.1). This reveals the influence that the electricity industry has exerted over policy formation by appealing to technical evidence and by claiming government validation of its position. It also demonstrates the relatively slight weight that inspectors are giving to public concerns on EMF, as represented by LPAs.

10.9.3 The role of planning officers

In the case study authorities, the drafting of policy dealing with the possible health effects of HVDT installations fell to planning officers (Figure 9.2). Although they took their own initiative in one or two cases (such as Tynedale (Section 7.4.3)), on the whole they were acting on the insistence of elected members; at Bridgend, for instance, "Members were very, very adamant ..." (Section 8.16.3). At Swindon, the local Friends of the Earth group played a role similar to that of members (Section 6.3). However, officers were also the recipients of representations made by the electricity industry, making them aware, firstly, of the industry's view and, secondly, of NRPB advice, which amounts to government backing for the industry's position. They also realised that other authorities were failing to bring EMF-related policies to adoption, because of inspector recommendations going against them (e.g. Section 8.12.2). This growing awareness led officers in a number of cases to recommend to their members that policies should not be based on EMF-related health fears, as there would be difficulties defending them at inquiry stage. These recommendations were accepted by the members of some authorities, such as Newham (Section 6.12.1). But elsewhere, members were adamant that policy should be pursued; at Rotherham, for example, "we were guided by our political masters ... they insisted that the policy stayed in" (Section 6.8.1). These examples are consistent with the figures for England and Wales as a whole, which show that LPAs had gone on to defend their policies at inquiry stage in about half of the cases where NG had brought an objection to "EMF" and "other development" policy concerns (Figure 5.5).

Officers have therefore occupied an uncomfortable position in relation to the expression of local EMF concerns. They have, in many cases, found that they themselves were required to defend proposed policies against their better judgement. This has, however, directed them into searching for means by which support might be lent to policies, both in the text of plans themselves and at public inquiry. Seeking to defend their council's position has, in fact, sometimes involved officers in considerable effort and determination and engaged them in searching out supportive information, including via the Internet (Sections 6.8.1; 8.12.2). This has led them into showing particular interest in precautionary approaches to policy.

10.9.4 Precautionary approaches

Several of the case study authorities made reference to taking a precautionary approach to EMF concerns (Table 9.3). For example, Swindon described its policy as "precautionary", following a suggestion of the Friends of the Earth (Section 6.3); Bridgend stated that it would assess proposals "with due reference to the precautionary principle" (Figure 8.12). Indeed, the Local Government Association (1999), in its response to the draft circular on land-use planning and EMF (Section 2.6.2) recommended the application of the precautionary principle in this context.

The precautionary principle

The precautionary principle embodies the idea that action should be taken in advance of scientific certainty in order to protect the environment (O'Riordan and Jordan 1995). It has considerable weight in the European Community, having been incorporated into the EC Treaty as a principle for action (Jans 1995); it has subsequently entered the UK planning guidance (DoE 1997c) and has been adopted as a principle of the government's strategy for sustainable development (DETR 1999c). However, it is interpreted to mean that there must be sufficient scientific evidence of possible harm (though not absolute proof) for precautionary action to be justified (Chadwick and Sienkiewicz 1998); measures must be based on "fact, not fantasy" (Her Majesty's Government 1994, p. 32).

The electricity industry, when objecting to EMF-based policies, has argued that any evidence of health risk is insufficient to warrant the application of a precautionary approach (e.g. Section 8.15). This reflects the understanding of the precautionary principle described above; indeed, the industry has been able to quote the government position in its defence (Section 6.7). Inspectors have generally accepted the industry's argument on this point (e.g. Calderdale Metropolitan Borough Council 1994).

The divergent standpoints on the implications of HVDT installations for residential areas therefore represent different understandings of the degree of uncertainty required before the precautionary principle should apply. This is illustrated by the Rotherham officer who expressed surprise at the rejection of his authority's approach: "... we felt that the jury is still out to an extent, and that until there was no doubt at all we should operate a precautionary principle" (Section 6.8.1). This mismatch between officer expectations and technical understandings exemplifies wider difficulties that are being experienced in implementing the precautionary principle within planning as a whole (Counsell 1999, Hales 2000) and within wider UK law (Bell and McGillivray 2000).

Prudent avoidance

Problems encountered in applying the precautionary principle within the context of EMF have led to consideration of other cautionary approaches that carry less stringent criteria. In particular, "prudent avoidance" is being advocated by some LPAs, in place of the precautionary principle (e.g. London Borough of Merton 2001). This has been defined as the adoption of "simple, easily achievable, low cost measures to reduce EMF exposure, *even in the absence of a demonstrable risk*" (World Health Organisation 2000; emphasis added). It was formulated as a policy measure specifically in relation to OHLs, and has been adopted as a general principle in some parts of the world since the late 1980s (UK Centre for Economic and Environmental Development 1991). Rotherham attempted to strengthen its development constraint policy by referring to the Canadian practice of distancing OHLs from dwellings (Figure 6.5). However, attempts to incorporate "prudent avoidance" into development plan policy are still likely to provoke objections from the electricity industry, which will continue to assert the lack of evidence of any significant risk; the notion also lacks any wider policy backing within the United Kingdom.

10.9.5 Wider policy frameworks

Apart from recourse to precautionary approaches, planning officers have had few sources of support for policy dealing with EMF concerns. Within the case study authorities, only one reference was made to hierarchical planning guidance in support of proposed policy: Rotherham appealed to guidance for the Yorkshire and Humberside region, which specifically stated that new OHLs should avoid residential areas and not pass over existing housing (Section 6.7). This is in some contrast to the "countryside protection" theme, where greater reliance was placed on national and regional planning guidance (Table 9.5; Section 10.7.2). This reiterates the lack of a wider policy framework that deals with EMF-related health concerns.

Indeed, some officers in the case study authorities specifically commented on the lack of clear government guidance on EMF-related planning issues: "I think this local authority, and presumably many others, would welcome some sort of statutory guidance" (Sections 6.4; 6.8.1). This implies that the NRPB is not always being recognised by LPAs as an authority on planning matters, despite being the government's statutory advisor on EMF issues (Section 2.6.2). This is partly due to a lack of awareness of NRPB advice; this is evidenced by officers dissuading members from pursuing policy once they *do* learn of the NRPB's position (e.g. Section 6.11). But the fact that some LPAs have persisted with certain policies even in the knowledge of the NRPB's advice suggests that it is not always being treated as an expression of the government's position. This may be due to the body's non-governmental status, and, possibly, it is perceived as being too closely aligned to the electricity industry.

There is room, therefore, for a definitive government statement on the planning of development in proximity to HVDT installations. Some inspectors have, in fact, called for the uncertainties raised by this policy area to be resolved at a national level (e.g. Lincoln City Council 1997). The draft circular on land-use planning and EMF produced in 1998 took the initial steps towards this, but progressed no further (Section 2.6.2). More recently, attention has turned instead to similar issues raised in connection with telecommunications installations, as shown by a revised guidance note on telecommunications (Department for Transport, Local Government and the Regions (DTLR) 2001b). It could be argued, given the continuing difficulties being faced by LPAs, that there is an equally strong case for equivalent guidance dealing with the planning of development close to HVDT installations.

There was also relatively little support for the "residential area protection" theme from strategies adopted for plans as a whole (Table 9.5). Rotherham's EMF concerns were consistent with the authority's overall health strategy (Section 6.8.1), and Aylesbury Vale's approach was in line with a wider theme of social accessibility (Section 8.12.2); otherwise, few explicit connections were made with broader plan aims. Similarly, there was a relatively weak level of policy history in relation to this theme (Table 9.5); of the case study authorities, only one had a direct precursor to its "residential area protection" policy (Section 8.11). This is in contrast to the "countryside protection" theme, which fitted into a stronger plan context, both in terms of the wider aims to which it related and its policy history (Section 10.7.2). This implies that the "residential area protection" theme (and EMF issues in particular) is a relatively new area of LPA concern, with few frames of reference within plan-making (Section 1.3). This, along with the difficulties experienced in adopting policy on this theme, may account for its less frequent occurrence across England and Wales as a whole (Figure 5.2).

10.9.6 Inter-authority contact

As noted above (Section 10.9.3), officers in many authorities have attempted to discourage the formation of policies that refer to EMF-related health concerns, following their realisation that other LPAs had faced difficulties in bringing such policies to adoption. This indicates a degree of inter-authority contact, especially at officer level. This was particularly evident in relation to some of the LPAs that did *not* pursue HVDT-related policy; for example, Newport County Borough

Council decided against pursuing policy based on EMF concerns after learning of a nearby authority's difficulties over the issue (Table 9.6).

However, inter-authority contact has also been a motivating force in favour of policy formation. "Residential area protection" is the only policy theme where there was a significant amount of communication between planning officers of different LPAs (Table 9.5). In some cases, communication was between neighbouring authorities. For example, informal consultation took place between the LPAs affected by NG's "North Yorkshire" line; this allowed Redcar & Cleveland to track the progress of a neighbour's policy through its plan preparation (Section 8.4.1). Also, Rotherham was encouraged in its pursuit of policy by similar approaches being taken by nearby authorities throughout south and west Yorkshire (Section 6.8.1). Figures 5.10 and 5.11 confirm these trends, by showing that there are particularly strong clusters of "EMF" and "other development" policy concern in these areas, with high levels of contestation between LPAs and NG. This degree of activity throughout the Yorkshire and the Humber region may also be associated with the regional guidance referred to above, suggesting that regional and local policy initiatives have together gathered a momentum in policy formation.

It is possible that longer-distance communication between officers has also been significant in encouraging policy formation. This is suggested by comments from one officer, in relation to Aylesbury Vale's different approach to development near OHLs (see Section 10.10.1): "We've had a lot of interest from other authorities who are aware that we're doing this … Some phoned us, via the grapevine" (Section 8.12.2). This kind of contact could contribute to the very dispersed pattern of policy occurrence also apparent in Figures 5.10 and 5.11, reflecting an informal officer network across the country.

10.9.7 Other policy instruments

Some of the case study authorities first expressed EMF concerns through other instruments, which then provided precursors for development plan policy. This was the case at Swindon, which initially developed its development constraint policy in the form of supplementary planning guidance(Section 6.3); similarly, local guidance preceded development plan policy at Redcar & Cleveland (Section 8.3). At Bridgend, a development brief for the authority's major housing site introduced the notion of development constraint along the length of OHLs (Section 8.16.3). The use of these alternative policy tools are more important in relation to "residential area protection" than to "countryside protection" (Table 9.5), because of the more site-specific focus of this policy theme, to which these instruments are more suited.

The preparation of these policy tools does not generally involve public consultation and so is not open to objections; indeed, it is unlikely that the electricity industry, for example, would be aware of references within them to HVDT installations. However, the non-statutory nature of these documents weakens their standing and makes them vulnerable to challenge in specific instances, such as at planning appeals; for instance, house builders objected to Bridgend's restrictions on development near OHLs (Section 8.16.3).

These instruments are relevant to development plan preparation because, firstly, they provide a source of ideas that can be transferred easily to proposed policies. For example, Redcar & Cleveland's local guidance provided much of the wording for the authority's proposed LP policy (Section 8.3). Secondly, they create a precedent, which justifies in itself the inclusion of certain topics in a plan; hence Swindon Friends of the Earth successfully argued that part of supplementary guidance should not be omitted from the statutory plan (Section 6.3).

10.9.8 Telecommunications equipment

A feature of some of the case study authorities raising the "residential area protection" theme is opposition to telecommunications equipment (Table 9.5). This has resulted from public anxieties about the possible health effects of exposure to EMFs emitted from mobile phone masts (or "base stations"). For example, at Bridgend, there have been strong protests against proposals for new equipment, and the council has decided to prevent the erection of base stations on council-owned schools and play areas (Section 8.16.3). Recently, public and member concerns have been directed more towards telecommunications equipment than HVDT installations (Sections 7.4.3; 8.8.2); this is also evident in LPAs with no HVDT-related policy (Section 9.8). Plans have reflected these concerns by including policies on the location of telecommunications equipment; Tynedale, for instance, introduced a policy paralleling many of the concerns expressed in its OHL policy (Section 7.4.3)

With regard to the formal procedures of plan-making, no reference was made to telecommunications equipment by case study authorities when forming HVDT-related policy; where the two matters have arisen, they have been treated as separate topics. Indeed, although the two types of installation raise similar planning issues, comparisons are not as valid as may initially appear. For instance, EMFs emitted by mobile phone base stations are of very different frequencies to those emitted by OHLs (Institution of Electrical Engineers 2001). However, there is evidence that the more recent public anxieties about the possible effects of base stations have reinforced existing concerns about OHLs. For example, at Mendip, campaigning against telecommunications equipment, headed by an elected member, seems to have increased sensitivity to the possible effects of HVDT installations, and to have contributed to the council's efforts to give policy expression to EMF concerns (Section 8.8.2).

10.9.9 Safety

Safety issues are associated with public health concerns in a number of the case study authorities. These are not always couched in precise terms; for example, Redcar & Cleveland simply implies that there are safety implications of OHLs in recreation areas (Figure 8.3). Swindon refers to the swing of conductors in adverse weather conditions (Section 6.3), while Rotherham mentions possible issues arising from vandalism (Section 6.7). In all cases, however, safety concerns are linked to other possible effects of HVDT installations and are used as justification for the same policy proposals. So Tynedale cites health and safety effects of OHLs as reasons for routeing OHLs away from residential and recreational areas (Section 7.3).

The electricity industry's response to safety concerns has been to refer to their own safety standards (Section 2.6.1); some modifications were made to policy wording as the result of these representations (e.g. Section 7.3). Safety issues are a relatively marginal element of the "residential area protection" theme and have perhaps been included as a means of bolstering more central health concerns (cf. "amenity" (Section 10.10)). The possible exception to this is in those LPAs mentioned above (Section 10.9.1), where a history of heavy industry has given the concept of "health and safety" a wide-ranging resonance. Nonetheless, it is difficult to imagine that safety concerns alone would have provoked policy formation in the case study authorities.

10.9.10 Hazardous installations

One of the case study authorities, Mendip, has sought to deal with health and safety considerations of OHLs under the heading of hazardous installations. Policies establishing safe distances between these installations and other development are understood to be potentially applicable to OHLs, though no specific references are made to them (Section 8.7). In taking this approach, the council accepts that possible health risk due to EMF exposure is not currently a planning consideration, but anticipates that this may change: "... if it can be demonstrated, that there is a health risk caused by power lines, we can use that policy in that context" (Section 8.8.2).

The difficulty with this approach is that regulations dealing with hazards focus not on installations which are sources of hazards, but on hazardous substances. These are chemical substances "which, by virtue of their toxicity, flammability or explosiveness, are capable of causing damage ..." (Miller and Fricker 1993, p. 177). Control consists of granting consent for the storage, use or production of substances of known danger on certain sites (under the Planning (Hazardous Substances) Regulations 1992). Although LPAs are encouraged to include policies in their development plans which ensure adequate distance between these sites and certain other land uses (DETR 1999a), the focus of attention is on minimising risk from accidental release of prescribed substances. It is difficult to envisage the extension of these provisions to non-chemical sources, such as electricity transmission, even if it is established that exposure to EMF is harmful to health. The approach taken by Mendip is not therefore likely to be adopted as a means of dealing with EMF issues; however, it does demonstrate the extent to which some LPAs are attempting to find ways of expressing EMF concerns in their plans.

10.9.11 Public concern

Several of the case study authorities specifically mention in their policies the level of public concern regarding EMF (Table 9.3). For example, Rotherham refers to "growing public concern" (Figure 6.5), and Bridgend, while accepting the lack of evidence linking EMF with ill health, states "there are genuinely held public perceptions to the contrary" (Figure 8.12); it is these *perceptions* that are being used in support of policy. Likewise, Mendip's strong emphasis on public participation in plan-making created a responsiveness to the public's views on the effects of OHLs (Section 8.8.3).

As discussed above (Section 10.9.2), objections to policies that refer to EMF have focused on the lack of scientific evidence to support health fears, and inspectors have generally accepted this argument. Moreover, they have sometimes stated that to take precautionary measures could unnecessarily increase public concern, especially among people already living close to OHLs (e.g. Swansea City and County 1998). This amounts to a dismissal of public perceptions when at odds with technical assessments of risk and of LPA attempts to represent public feeling. This is consistent with legal rulings on similar matters. For instance, objections to a proposed incinerator, on the basis of public apprehensions regarding emissions, were effectively dismissed, with the Court of Appeal judge stating, "Public concern is ... a material consideration ... But if in the end that public concern is not justified, it cannot be conclusive" (Gateshead MBC v Secretary of State for the Environment (1995) Env LR 37).

However, greater acknowledgement was made of the weight that might be attached to public environmental concerns in another case, where it was judged that public perception of risk, even if unsubstantiated, could be a material planning consideration (Newport BC v Secretary of State for Wales and Browning Ferris Ltd (1998) Env LR 174). This approach was acknowledged in the draft EMF circular (Section 2.6.2). Inspectors at some recent planning appeals concerning telecommunications equipment have indeed attempted to place greater weight on public perceptions of harm (Wood 2001).

The question of how much account planning decisions should take of public perceptions of harm, as opposed to scientifically established criteria, therefore remains a vexed one, particularly for LPAs who inevitably find themselves representing the interests of their localities. The electricity industry's approach to this issue has generally been to try to allay fears through disseminating information about scientific findings (Section 2.10.2). It is possible that there is potential for greater initiatives to be taken in seeking to increase public confidence in the results of current research.

10.10 AMENITY

10.10.1 Expression of amenity concerns

Along with health and safety concerns, the "residential area protection" theme is characterised by concern for other possible effects of HVDT installations in residential areas (Table 9.5). For example, Aylesbury Vale states that OHLs "are visually intrusive, the lines are often noisy and the electromagnetic fields that surround them can cause radio interference" (Figure 8.9). In the early stages of plans, these effects are generally twinned with health and safety concerns, and thus provide additional justification for proposed policy measures. So Tynedale states, in its deposit draft plan: "The visual impact and concern over ... the health and well-being of residents ... makes it desirable to route lines away from residential areas" (Figure 7.1). It seems unlikely that non-health-related issues would have been sufficient in themselves to provoke policy; it is more probable that, as with safety matters (Section 10.9.9), they were raised as a means of strengthening health concerns.

In several cases, non-health effects were referred to as affecting the "amenity" of residents (e.g. Figure 6.5). This draws on the notion of "residential amenity" (a phrase specifically used by Bridgend (Figure 8.12)), which tends to concentrate on the design and layout of housing and surrounding development, in order to make for better living conditions (Millichap 1995); this is a more urban ideal than the conservationist concerns of the Electricity Act's Amenity Clause (Section 2.3.3).

Moreover, amenity is a theme that was used by those authorities that had difficulties with policies based on EMF fears (Section 10.9.2). For example, Newham changed the justification for one of its proposed policies from health concerns to "visual amenity", following an objection from NG (Section 6.11). Similarly, the inspector of the inquiry into Swindon's plan recommended that the "general amenity of dwellings" should be the council's only consideration (Section 6.3). At Aylesbury Vale, the council pre-empted these difficulties by referring only to amenity reasons in its proposed policy and adopting the term "amenity zones", despite the fact that EMF fears had clearly provided impetus for the policy; moreover, it sees itself as providing a model for other authorities to follow in this respect (Section 8.12.2).

The expression of amenity concerns therefore represents not only interest in specific non-health-related effects of HVDT installations in residential areas, but is also a fall-back position from the problems encountered in trying to give policy statement to health fears. The question arises, however, are LPAs now relying solely upon strict amenity matters in order to justify certain policy measures, or is "amenity" being understood to embrace, albeit covertly, EMF concerns? There is, at least, an ambiguity now arising in the use of the term, and it is possible that certain actors are beginning to invest it with new meaning in this context. The historic lack of definition, yet evolving applicability, of "amenity" (Cullingworth and Nadin 2002) makes this a likely prospect.

10.10.2 Amenity effects

In summary, a number of non-health effects are described as affecting "amenity" (Table 9.5). Visual effects are the most frequently referred to; for example, Rotherham states that views from properties near OHLs could be "significantly impaired" (Figure 6.5). Noise and possible radio interference, and so on, are also occasionally mentioned. The possibility of these effects was not challenged by the electricity industry and accords with NG planning guidelines (Section 2.10.2), suggesting that there is a common understanding of some of the issues raised by HVDT installations in residential areas.

Cumulative effects

In addition to these individual effects, the case study authorities also hint at cumulative effects of OHLs; their overall presence is such that they are considered dominating and oppressive when located near housing (Figures 2.1, 6.5 and 6.10; Section 8.12.2). It is possible that this language touches again on the perceived health implications of OHLs. It also raises questions about the symbolism of OHLs; how are these large-scale industrial structures interpreted when in close proximity to people's dwellings? To what extent do their physical characteristics generate the kind of anxieties discussed above? These questions, though beyond the remit of the electricity industry documents reviewed in Chapter 2, merit further attention by those concerned with the juxtaposition of HVDT installations and residential areas. Although a number of North American studies of perceptions of OHLs were carried out throughout the 1970s and 1980s (Furby *et al.* 1988), there is potential for further research in this area, particularly within the UK context.

Property values

The case study authorities also make some reference to the possible effect of OHLs on the market value of nearby land and properties (Table 9.3; Figures 8.9 and 8.12). This approach has been challenged as not dealing with a land-use planning matter (Section 8.11); this view is supported by government guidance (DoE 1997a). Whether property values are significantly affected by the nearby presence of OHLs is uncertain; again, this is an area that has been part of North American rather than UK studies (Priestley and Evans 1996). There is, therefore, scope for research into this issue, though, from a planning perspective, this would be better placed within a wider context of the social effects of HVDT installations in residential areas.

10.11 POLICY MEASURES

Several policy measures have been put forward to protect sensitive residential areas from the potential effects of HVDT installations (Table 9.5); these are usually designed to maintain distance between urban development and installations. Firstly, there is outright resistance to new installations near residential areas, such as by Mendip (Figure 8.5). Secondly, development constraint corridors are set along the length of lines, such as by Swindon, which sets out a 20 m "exclusion zone" within which no housing is allowed (Figure 6.1). Thirdly, some consideration is given to the form of development near OHLs: Bridgend states that the impact of OHLs can "be mitigated in part by the design and layout of new residential development" (Figure 8.12). Finally, placing lines underground is suggested as a means of alleviating the possible effects of OHLs in residential areas, such as by Tynedale (Figure 7.1).

Planning experience

These measures have been put forward partly as a result of planning experience (Table 9.5). Officers in a number of authorities have been informed by particular cases in their areas, some of which have provided specific policy precedents (Figure 9.2). For example, Aylesbury Vale's policy of prohibiting residential development within 50 m of OHLs was explicitly based upon a local instance of development control in which a similar restriction had been enforced (Section 8.12.2). Bridgend's slightly less prescriptive approach also reflects the experience of development in the vicinity of OHLs (Section 8.16.3; Figures 8.14 and 8.15). Instances of dealing with the juxtaposition of HVDT installations and residential areas have therefore presented ideas that have been incorporated into more widely applicable policy. The site-specific nature of this policy theme inclines it particularly towards this kind of interaction between development control and forward planning.

Objections to policy measures

As discussed above (Section 10.9.2), the electricity industry has brought strong objections against the measures proposed to maintain distance between residential development and HVDT installations, by rejecting the notion that EMF exposure presents a health risk. At a policy level, this can be seen as straightforward adherence to official advice on EMF exposure, as stated by the NRPB. The industry's determined stance on the issue has, however, raised suspicions about underlying motives that any acknowledgement of EMF-related health risk could pave the way for legal claims for compensation from people already living close to OHLs (Section 6.8.1).

Another possible reason for the industry's resistance to the establishment of development-free corridors alongside OHLs relates to land values. In general, utilities do not own the land crossed by OHLs, but instead negotiate "wayleaves" (licences providing the right to install an OHL) with landowners (NG 1999c). The industry may feel that landowners would be reluctant to agree to wayleaves if, in so doing, a large corridor of their land were effectively sterilised from future development. Additionally, the industry may be seeking to limit compensatory payments to landowners in this position (it is, in fact, industry practice to offer limited compensation where development potential is restricted by the presence of OHLs.)

However, it should also be noted that, from the electricity industry's point of view, it is preferable not to have built development in close proximity to OHLs, as this can lead to serious difficulties of access and maintenance. Thus, NG calls on developers to give consideration to amenity when building near existing OHLs (Section 2.10.2). There is, therefore, a degree of ambiguity in the industry's position of opposing policies that seek to distance development from OHLs. The fact that it does so with considerable determination suggests that the weight of argument falls clearly on the side of avoiding the possible consequences of set-width corridors becoming common practice, particularly when established out of concern for potential health risk.

Some objections were also brought against proposals to establish development-free corridors alongside OHLs by the house-building industry (Table 9.5), arguing against the implied danger to health (Sections 6.3; 6.7). However, house builders have also referred to the loss of development potential that would result from these restrictive corridors (Section 8.11). Clearly, it is in the interest of developers to make the greatest possible use of a site. Moreover, the experience of Bridgend suggests that the presence of OHLs is not a major deterrent to house builders; although they may show initial reluctance to take on sites affected by OHLs, once they do so, they will build housing as close to OHLs as is permitted (Section 8.16.3). However, discussion on these policy measures has been so dominated by the issue of possible health risk that little consideration has been given, in inspectors' reports, for example, to other questions of the best use of land underneath and alongside OHLs. There is potential, therefore, for more detailed consideration to be given to the design of housing, and other appropriate types of development and land use, close to OHLs.

ENCOURAGING REGENERATION

Table 9.5 and Figure 9.3 show the components of the "encouraging regeneration" theme and form the basis for the headings given in the following text. Again, areas that are considered sensitive to HVDT installations are considered first. This is followed by an examination of policy content, both the perceived effects of installations in those areas and the measures proposed to offset them. The policy framework within which this theme operates and the role of different actors in policy development are then discussed.

Regeneration issues were central to only one of the case studies, Newham; this case provided the basis for identifying "encouraging regeneration" as a distinct theme within HVDT-related policy (Section 9.5). However, questions of regeneration were also evident in other case studies, even though they were not made explicit in the wording of policy. The following discussion therefore also makes reference to them.

10.12 REGENERATION POTENTIAL AND HVDT INSTALLATIONS

10.12.1 Sites with regeneration potential

As in the previous two themes, this policy theme is characterised by areas that are considered vulnerable to HVDT installations (Table 9.5). Ironically, these are of low quality – sites of dereliction, rather than scenic countryside or residential areas. Newham, for example, has large degraded areas, resulting from industrial decline and the closure of the docks: "miles and miles of wasteland" (Section 6.12.2). Rotherham and Redcar & Cleveland also have sizeable areas of dereliction due to the decline of heavy industry (Sections 6.6; 8.2).

Some of these locations are also affected by existing HVDT installations. Major industrial zones were originally attractive locations for HVDT installations, partly because they were heavy consumers of electricity and also because they provided convenient corridors for OHLs into urban areas (Sections 6.8.3; 6.12.2). However, these installations have remained in place despite the loss of industry, and are still operational as they continue to supply other areas. The result is that they now form the surviving relics of industrial landscapes and have become surrounded by degradation and abandonment.

The sensitivity of these areas to HVDT installations arises from their perceived potential for economic regeneration. HVDT installations, especially OHLs, are seen as a threat to future development, which is generally envisaged to be of a very different nature to that existing previously. This leads to a call for OHLs to be re-routed away from these areas or placed underground (Figure 6.9). In consequence, policy tends to be site specific, referring to precise locations and particular OHLs.

10.12.2 Policy content

This policy theme is concerned with two effects of HVDT installations: visual intrusion and occupation of land (Table 9.5). Visual intrusion is described in general terms; for example, Newham refers to one OHL as a "visual and environmental blight" (Section 6.12.2), suggesting an overall sense of degradation of the surroundings, as with "amenity" effects (Section 10.10.2). Occupation of land is seen as a physical constraint on development (Figure 6.9). Surprisingly, only this policy theme raises the issue of land-take by HVDT infrastructure (Table 9.5); this is in some contrast to electricity industry practice, where it is considered to be an important issue (Table 2.2).

Despite the assertion that OHLs inhibit regeneration, no suggestions are given about the types of development that are being deterred or how serious an effect is being imposed. It appears that the policy stance reflects an impression of the negative influence of HVDT installations rather than any analysis of the degree to which they actually act as a deterrent to developers.

The policy measures proposed to overcome the perceived negative effects of HVDT installations are radical: the complete re-routeingor undergrounding of existing, major OHLs (Table 9.5). Newham proposes to identify corridors in the borough where lines could be placed to replace the existing OHLs (Figure 6.9); this idea is developed further in more recent development proposals, in which corridors for all major services are suggested (Section 6.12.2).

Electricity industry objections to the proposed measures were based primarily on the great difficulties that would be involved in carrying them out (Section 6.11); undergrounding especially of high-voltage lines is prohibitively expensive and has significant environmental consequences (Section 10.6.2). Inspectors have shown some support for the council's position, but have not fully endorsed their proposals (Sections 6.11; 6.12.2). As in the "countryside protection" theme, this theme demonstrates a lack of knowledge on the part of LPAs of the practicalities that would be involved in implementing their proposed measures.

Of the policy concerns throughout England and Wales, "visual effects" most closely represents this policy theme. Figure 5.7 shows the distribution of this policy concern, but does not distinguish between the visual effects expressed in different policy themes. However, it does show a cluster of DPPs around the Thames Estuary, which is likely to reflect regeneration issues such as those at Newham. It is possible that visual concerns expressed in other old industrial areas, such as Teesside and south Yorkshire, also reflect the "encouraging regeneration" theme.

10.12.3 Policy framework

This policy theme operates within a strong policy framework (Table 9.5). Newham places its OHL policies firstly within the context of its plan as a whole, which sets out an ambitious regeneration strategy (Section 6.10). It also refers extensively to regional guidance, which describes the borough as being in a "pivotal position", and specifically mentions OHLs as being detrimental to regeneration (Section 6.12.2); a similar hierarchical framework exists for Rotherham (Section 6.6). At a more local level, Newham also has a history of this policy theme, and supplementary planning guidance for locations affected by OHLs continues to press the issue (Section 6.12.2).

The overall strategy of regeneration is strongly linked within this policy framework to the notion of environmental improvement. For instance, Redcar & Cleveland associates environmental protection measures with improving the borough's image, and thus encouraging regeneration (Section 8.2). Newham's regional guidance speaks of setting new environmental standards for the area, in order to break out of a "cycle of environmental blight" and puts forward the undergrounding of OHLs as an example of achieving this; again, this aims to offset the conditions that inhibit regeneration (Section 6.11; 6.12.2).

This policy theme is therefore being driven by strong economic imperatives, which have captured environmental objectives as a means of achieving the goal of regeneration. Although Newham is the only case study authority to connect these aims formally to HVDT-related policy, similar trends are evident elsewhere; for example, at Redcar & Cleveland, there was some feeling that the presence of OHLs had adverse economic effects (Section 8.4.2).

10.12.4 Actor involvement

This policy theme gathered strong support from both planning officers and elected members at Newham and also from a development corporation in place at the time of drafting policy (Table 9.5). This demonstrates a consensus around the regeneration strategy, reflecting both the scale of dereliction within the borough and a shared vision of what could be achieved. Their determination to overcome the perceived problems presented by a key OHL is evident from the Wayleave Hearing that was provoked by their action (Section 6.12.2). It came to be considered "essential" to regeneration for a major OHL to be moved or placed underground (Figure 6.9).

Subsequent experience proved that the OHL in question did not act as the disincentive to development that had been feared; this is consistent with the experience of house builders, as discussed under the "residential area protection" theme (Section 10.11). Nonetheless, the view persists that OHLs elsewhere in Newham could impede progress (Section 6.12.2). Moreover, regional concerns that OHLs are decisively deterring investment remain; a recent article entitled *Pull Lines Down and Watch New Schemes Rise* states that OHLs are making the "Thames Gateway effectively undevelopable" (Hall 2002). But this claim is unsubstantiated and is not supported by the experience of Newham. However, it is probable that Newham has circumstances working in favour of regeneration that are not shared by other authorities along the Thames Estuary, such as its proximity to central London. There is scope for further study, therefore, into the degree to which the presence of HVDT installations, and other geographical factors, are working against the regeneration of different areas.

CHAPTER 11

Conclusions

11.1 A GENERALISED PROCESS OF HVDT-RELATED POLICY FORMATION

The approach adopted in this book has assumed that policy formation incorporates the influence of a wide range of features both external and internal to policy-making, the role of different actors and the procedures of plan-making itself. Widespread patterns and forces, and local contexts and actor involvement, have been understood to determine policy (Section 4.2.1; Healey 1997a). These have been explored by means of an analysis of the components of policy formation and by generalising from these components to wider overall processes (Section 4.4). This approach has produced a comprehensive description of the parallel themes evident within HVDT-related policy, each dealing with distinct issues and concerns and common processes occurring within them (Chapter 10).

However, it is clear that there are areas of overlap between the themes identified and that certain influences contribute to more than one of them. Moreover, they are often treated simultaneously within development plans, frequently under the same heading, and with a degree of mutual reinforcement. HVDT-related policy has, therefore, an overall dimension, in addition to the expression of specific themes; certain processes are at work in raising general concerns and in bringing policy as a whole to expression. This overarching level of policy formation can be described by generalising from the individual themes set out in Chapter 10.

11.1.1 Elements of Policy Formation

The following list describes the main elements at work in policy formation, discussed in Chapter 10, as they apply across HVDT-related policy as a whole:

1. Local planning authorities (LPAs) identify specific areas that they consider sensitive, in one way or another, to the actual or possible effects of HVDT installations.

High Voltage Electricity Installations S. Jay

2. The identification of these areas is, in part, an expression of local strategies (e.g. environmental, housing, regeneration) that increase the sense of value attached to them.
3. Existing HVDT installations are perceived as affecting, or potentially affecting, sensitive areas.
4. Recent or potential HVDT projects are often seen as threatening sensitive areas.
5. There is often the prospect of an increase in the perceived effects of HVDT installations, resulting from wider economic, social, and so on, trends (e.g. energy policy, development pressure).
6. National public debates and environmental concerns relating to HVDT installations heighten local concerns.
7. Hierarchical policy frameworks at national, regional and county level, and local planning strategies provide support for HVDT-related policy.
8. Local planning cases and planning experience of HVDT installations generate policy concerns and provide policy precedents.
9. Other local planning instruments are sometimes used to express HVDT-related policy and provide a background for development plan policy.
10. Inter-authority contact sometimes plays an important role.
11. A history of local HVDT-related policy sometimes provides momentum.
12. Planning officers have a central role in formulating and developing policy, partly in response to proposals and subsequent objections.
13. Elected members play an initiating role, sometimes holding strongly to their positions.
14. Interest and environmental bodies and institutions play a supportive, and sometimes initiating, role.
15. The public plays a background, supportive role.
16. The electricity industry and some other bodies show a readiness to oppose policy that they consider detrimental to their interests.
17. Inspectors tend to favour arguments put by the electricity industry and to recommend that LPAs should qualify their opposition to HVDT installations and their proposed measures.

These elements of policy formation clearly operate in conjunction with each other, such that policy comes to expression as they accumulate. The degree to which an LPA pursues its concerns in the face of objections may also reflect the extent to which a range of different influences are at work. The cumulative nature of these components has been most apparent in the northeast of England, where the effects of existing HVDT installations in sensitive areas, a highly controversial transmission project, inter-authority contact and other features have given rise to a cluster of LPAs pursuing several policy concerns, with high levels of contestation with the electricity industry. It is likely that other areas where there is intense policy formation are also characterised by the convergence of a number of the elements listed above (e.g. Sections 6.8, 7.8).

11.1.2 Processes in Policy Formation

The links between the elements of policy formation can also be described for HVDT-related policy as a whole. Figure 11.1 is a network diagram for the overall process of policy formation; it is based on the equivalent diagrams for the three themes (Figures 9.1, 9.2 and 9.3) and portrays a synthesis of the main elements of these figures. It shows that there is a consistent pattern of the influence of certain features on actors who initiate policy and of policy modification during development plan preparation.

The elements shown in Figure 11.1 can be further simplified into key groups of components, as indicated in Figure 11.2; this suggests that HVDT-related policy arises from and is shaped by four interacting processes:

1. *Sensitisation to HVDT installations.* The existing and potential presence of HVDT installations in areas that are being attributed an increased sense of value creates a perception of actual or possible intrusion in those areas. Media influence may play a part in raising concerns.
2. *Policy initiation.* Sensitivities to the effects of HVDT installations feed through to policy makers, both directly and via the public and/or interest groups. Members may mediate wider concerns to officers. The local experience of HVDT cases, resulting from recent or current projects, may contribute policy concerns and ideas.
3. *Policy framework.* Support and ideas for policy initiatives may be taken from existing local and hierarchical policy, nearby authorities and other institutions. Broad aims of the plan may also reinforce HVDT-related policy. But the wider policy framework may also present difficulties to policy makers.
4. *Policy contestation.* Consultation on proposed policies may draw support from interest/environmental bodies and provoke opposition from the electricity industry and others. The LPA may defend its approach at the inquiry/examination stage, but is likely to have to qualify its stance on HVDT installations.

The formation of HVDT-related policy therefore originates primarily with the increased sensitisation of certain areas to HVDT installations. This itself arises from the convergence of two opposing land-use issues: the attribution, for a variety of reasons, of enhanced environmental value to certain areas, and the presence, or likely future incursion, of HVDT installations. The way in which this process operates varies greatly according to local geographies and backgrounds, but occurs in one form or another throughout the body of HVDT-related policy.

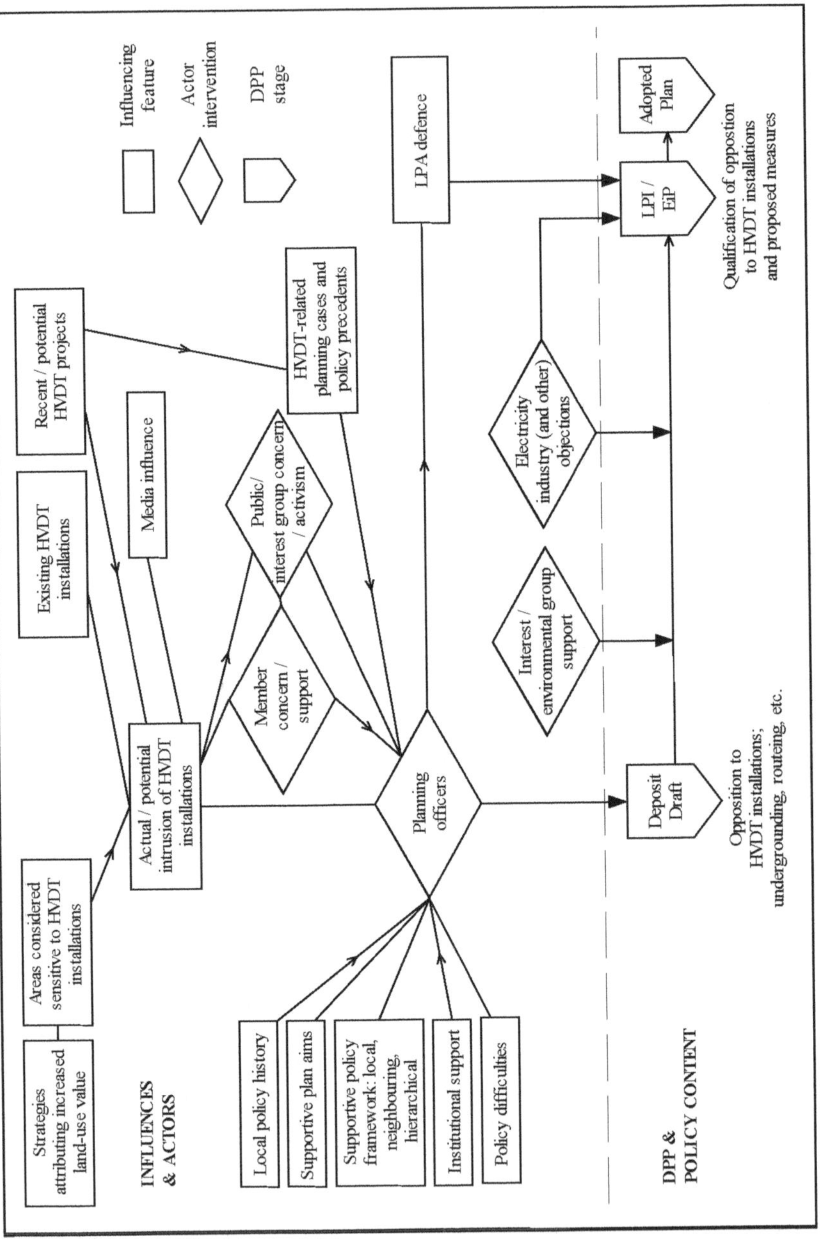

Figure 11.1 Generalised process of HVDT-related policy formation

The attribution of value to certain areas is clearly expressed in the wording of HVDT-related policy, in references to the protection of designated areas, amenity interests, health concerns, and so on. It is also implicit in connections that are made to the wider environmental aims of plans, such as the conservation of environmental assets, and environmental improvement in the interests of regeneration. This links HVDT-related policy directly to the environmental/sustainability strategies that have taken an increasingly central place in plan-making for reasons discussed previously (Section 3.3.4). This suggests that, despite the difficulties being encountered in the interpretation of these strategies (Owens and Cowell 2002), they nonetheless demonstrate new attitudes towards places and are seeking to invest areas with a new, possibly "post-material", sense of value (Owens 1994 p. 445). Certainly, HVDT-related policy is an explicit attempt by LPAs to enhance the quality of local environments such that HVDT installations are now regarded as degrading intrusions within them. The patterns and trends of installations that contribute to this sense of environmental damage have already been discussed (Sections 10.3, 10.4, 10.8); future possible trends are further considered in Section 11.3.1.

Policy initiation follows on as a response to heightened awareness of the intrusion, or possible intrusion, of HVDT installations in particular locations. It is at this point that those actors most concerned about the perceived effects of installations are prominent, giving rise, directly or indirectly, to draft policy. The wider policy framework is also instrumental in both instigating and reinforcing local policy, as is experience gained through the local planning of HVDT installations. Clearly, the preparation of development plans in itself can be seen here as a factor in policy formation; given the enhanced status of plans since 1991 (Section 3.3.6), they have provided an obvious means by which LPAs have chosen to mediate their concerns and seek to offset the unwanted effects of HVDT installations.

The procedures for development plan preparation allow for draft policy to be supported and challenged; given the stance generally taken by LPAs, policy initiation is typically followed by policy contestation, most usually as a result of objections by the electricity industry. This marks a progression from the technical drafting of policy by officers, through consultation and negotiation with objectors, to an increasingly adversarial approach, possibly culminating in a semi-judicial inquiry (Adams 1994). LPAs look again at this point to wider policy frameworks that support their positions (Section 10.7.2). But inspectors are likely to favour arguments advanced by the electricity industry, demonstrating a tendency to support prevailing structures (Section 3.6.1; Owens 1997b).

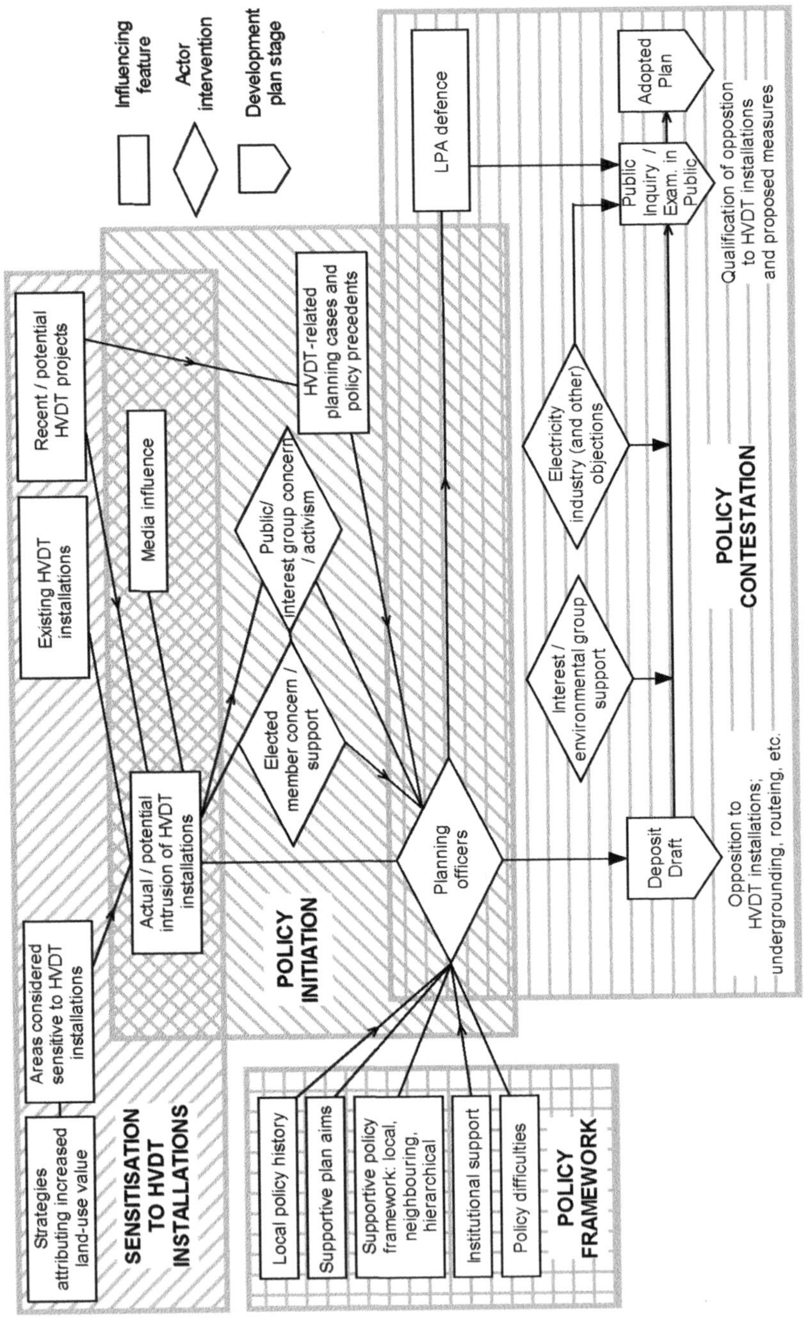

Figure 11.2 The overall formation of HVDT-related policy

11.2 THE PROTECTION OF LOCAL ENVIRONMENTAL QUALITY

Common to the three policy themes, and an overarching principle of HVDT-related policy, is the aim of protecting local environmental quality from the perceived effects of HVDT installations. This implies a prioritising of local interest over and against wider interests; although this is not explicitly stated in HVDT-related policy, LPAs are demonstrating the greater regard they give to the environmental conditions of their localities than to the wider benefits associated with electricity supply.

The significance of this assertion can be seen in relation to changing local authority attitudes towards the electricity companies as utilities. Development plans typically include "utilities" policies that facilitate the provision of services, both locally and also as a contribution to wider need; this is in line with government guidance encouraging LPAs to consider the requirements of the utilities for land (DETR 1999a). The fact that HVDT-related policy is now concerned primarily with environmental quality, with very few references to conditions under which HVDT infrastructure would be acceptable, suggests a marked change of attitude on the part of LPAs towards this sector. It suggests that the notion of "public utility", by which the development of HVDT systems has in the past been justified (Section 1.2), is not currently a strong consideration for LPAs. Local interests are now being favoured over the needs of the electricity industry, even in preference to a wider public good; the readiness of many LPAs to defend policy against representations from the industry (Section 5.4.2) is an indication of the degree to which they seek to assert these immediate interests.

It is probable that this shift in emphasis is linked to the privatisation of the electricity industry. The privatised electricity companies have been seen by the public primarily as business concerns rather than as service providers, with shareholders benefiting at the expense of consumers (Surrey 1996). Likewise, LPAs may have come to view them more as private developers than as statutory undertakers and have less sense of responsibility towards providing for their needs. Certainly, LPAs have been aware of an increasingly market-led approach to electricity generation, resulting in the construction of new power stations and grid connections. This may make LPAs feel that they should take a stronger role in regulating the industry (Section 8.15) and that the industry should bear the environmental costs of its activities, by placing lines underground, for example. It is certainly likely to encourage LPAs to take a stronger role in defending the interests of their areas and their residents in the face of the undesirable effects of HVDT operations. Although the neo-liberal agenda, of which privatisation of the utilities was a part, generally required planning to ease restrictions on development (Allmendinger and Thomas 1998), utilities have, ironically, found themselves in a less favoured position by being thrust into the private domain.

This shifting attitude on the part of LPAs runs parallel to broader changes in the public perceptions of HVDT installations, such that former acceptance of installations as necessary for electricity provision has given way to regarding them as unwelcome intrusions on the landscape (Section 10.10.2). It is now expected that higher environmental standards should be set, and that services should be provided with less damaging consequences. This is most apparent in relation to anxieties about electromagnetic field (EMF) exposure, but also in relation to the effects of HVDT installations on the rural landscape; hence the expectation, for example, that overhead lines (OHLs) should avoid sensitive areas or be placed underground along with other services (Section 10.6).

Along with this lessening of a sense of responsibility towards statutory undertakers, LPAs are also seeking to counter, or avoid, a disproportionate presence of HVDT installations in their areas. Some LPAs clearly feel a sense of disbenefit arising either from a high concentration of installations in their areas, serving major industry, or because they see themselves as corridors for the export of energy generated in the north and midlands to the south of England (e.g. Sections 6.12.2, 8.4.1). In such cases, HVDT-related policy is an expression of the interests of one locality over those of others or of local versus national interest. HVDT installations have features in common with certain other “locally undesirable land uses”, in that they bring few immediate benefits to the areas where they are situated, but have significant adverse consequences. This puts LPAs in the position of resisting what they perceive to be an unfair share of infrastructure, even when they accept a responsibility to contribute towards national need. This exemplifies a wider growing unwillingness on the part of communities and their representative authorities to play host to activities that are considered essential nationally or regionally, but which have serious environmental effects locally (Scott *et al.* 1998); in this respect, HVDT demonstrates the planning difficulties faced by many other public utilities, such as transport infrastructure and waste disposal.

By placing emphasis on the protection of local environmental quality, LPAs are, therefore, asserting a less compliant relationship with the privatised utilities and setting limits on their willingness to accommodate regional/national infrastructure. The defence of the environmental conditions of their areas (often in support of local economic and social aims) has taken priority over the wider interests represented by the presence of HVDT installations.

11. 3 FUTURE DIRECTIONS IN HVDT-RELATED POLICY FORMATION

There are a number of current trends and uncertainties in relation to the energy sector and to planning practice that are likely to have consequences for HVDT-related policy. These are explored below, in order to indicate the possible future development of policy.

11.3.1 Trends in Electricity Generation

Scenarios for future electricity need in the United Kingdom suggest that demand will continue to increase at a rate similar to the present rate (NG 1999a); pressure for new generation capacity will therefore remain. This is likely to result in continuing applications for new generating plant, with the associated need for new grid connections. The most recent recommendations for meeting demand include increased use of gas-fired power stations, with the options of new nuclear and clean coal generation being kept open. The expanding role of the renewables sector is also recognised, with an increased target of 20% of total generation from renewable sources by 2020 being targeted (DTI 2003). The consequences that they would have for HVDT installations are uncertain. However, the overall trends in generation are such that applications for transmission lines to major plant, and for distribution lines to "embedded" smaller-scale renewable projects, are likely to be significant in number (Royal Commission on Environmental Pollution (RCEP) 2000). It is likely that some renewable schemes will require major grid connections, such as large-scale wind farms in upland Britain and offshore (DTI 2002); these raise the prospect of major associated HVDT installations. Smaller renewable projects are likely to be numerous, causing a large number of applications for lower-voltage lines. It is probable, therefore, that an increasing number of LPAs will have direct experience of applications for HVDT installations and will express policy concern over the prospect of future applications; these are likely to include LPAs in areas where policy concern is not currently strong.

These trends in the nature of electricity projects raise questions about the relative location of generation and consumption. There have been suggestions, for example, that bringing generation closer to demand, and moving towards smaller-scale, localised generation, would reduce the impact of grid connections (Patterson 1999). Indeed, Ofgem, the industry's regulator, has indicated the possibility of providing greater incentives for supply and demand to be located closer to each other (Ofgem 2001). Siting generating plant close to consumers rather than the energy source is, however, likely to be a long-term goal, with little immediate benefit regarding the effects of HVDT installations; it may be, though, that some LPAs will press for this through their plans (Section 10.6.3).

11.3.2 The Electricity Industry's Environmental Practice

The increased attention that has been given by the electricity industry to the environmental effects of its activities is likely to continue. This is evident from the recent production of company environmental reports, planning guidelines, and so on, and the efforts to find technical means of minimising effects (Sections 2.8–2.11). This has been in response to more stringent environmental regulations upon the industry and has also resulted from the experience of planning difficulties of the kind described in this book. However, the extent to which the industry is prepared to give greater consideration to the environmental effects of HVDT installations remains open to question. The privatised status of the industry does not encourage costly action beyond that which is required by regulations, and the current regulatory oversight of the industry does not allow electricity companies to transfer the costs of further environmental protection to customers. However, Ofgem has recently suggested that, as a regulatory body, it should be able to place greater emphasis on environmental considerations by setting environmental targets for the industry which take into account certain externalities (Ofgem 2001). It is likely that this would encourage further improvements in practice and would subsequently give greater reassurance to LPAs that the electricity industry was operating at better environmental standards.

11.3.3 The Future of Plan-making

There are a number of possible directions with regard to the future of development plan preparation. Firstly, current reform of the planning system has cast doubts on the future shape of plan-making (DTLR 2001a). For example, the level of detail that plans should incorporate is questioned; the emphasis is upon shorter, simpler plans (Section 3.3.7), which may lead to the exclusion of policies on comparatively minor issues such as HVDT. Set against this is the requirement for action plans, focusing on specific areas, which could encourage greater policy attention to be given to the effects of HVDT installations within those areas. This does not respond fully to the recommendations of the RCEP (2002), which calls for a more strategic and comprehensive approach to plan-making, with a strong emphasis on safeguarding the environment (Section 3.3.7). Nonetheless, at a regional level, the new "regional spatial strategies" could allow a move in this direction with energy infrastructure included within their remit, thus allowing a more strategic consideration of HVDT-related issues.

Secondly, the growing emphasis on environmental/sustainability objectives within plan-making is likely to be maintained. The government has continued to assert its commitment to sustainability principles in general (DETR 1999c), and within plan-making in particular (DETR 1998). Although the degree to which the concept of sustainability has been effective in plan-making is open to question, it now has a strong "rhetorical grip" within planning practice (Owens and Cowell 2002, p. 26). The place of the strategies that have encouraged the formation of HVDT-related policy therefore seems assured, and they are likely to continue to operate in a similar way.

The importance of environmental considerations in development plans is, in fact, likely to be strengthened by recent European legislation, which now requires a form of strategic environmental assessment to be carried out on land-use plans (Directive 2001/42/EC). It is not yet clear exactly how this requirement will be implemented in relation to local authority development plans and how it will affect the content of plans. However, the assessment of plans is likely to give greater weight to the kind of environmental considerations that have been central to the formation of HVDT-related policy and may well favour the inclusion of policies seeking to protect areas from the effects of HVDT installations.

11.4 RECOMMENDATIONS

HVDT-related policy has been shown to be highly contested in certain respects and subject to a considerable number of objections, especially from the electricity industry (Section 5.4.2). This is indicative of conflicting attitudes towards HVDT installations and also of differing understandings of the effects of HVDT installations and possible means of offsetting them (e.g. Section 10.6.2, 10.9.2). There are also indications, however, of willingness by different parties to accommodate one another's concerns to a greater or lesser extent (e.g. Sections 10.2, 10.9.3). This suggests that there is scope for a more consensual approach to the planning of HVDT-related development and to the formation of policy in which "participants learn from each other not merely about facts, interests and preferences, but about what participants care about and why" (Healey 1997b, p. 267). The following recommendations aim to facilitate this approach and also to encourage improved practice in the planning of HVDT-related development.

1. Forums enabling communication

Most exchanges of differing perspectives on HVDT-related planning issues currently take place through the adversarial procedures of plan-making. Opportunities should be created for more collaborative discussion, either initiated by individual parties or arranged cooperatively:

- The Planning Officers' Society or the Royal Town Planning Institute could organise forums for HVDT-related planning matters, consisting of representatives of environmental and interest groups, the electricity industry and LPAs.
- LPAs could hold round-table discussions with representatives of the electricity industry and local environmental and interest groups, at the early stages of preparing HVDT-related policy.
- Electricity companies could involve other bodies more fully when preparing their Schedule 9 statements (Section 2.3.3), and extend the list of consultees beyond that which is statutorily required, to include representatives of LPAs and interest groups.

2. Strategic planning

In view of the likely expansion of certain sectors of the energy industry (Section 11.3.1), a strategic approach should be taken to the siting of generating plant and associated connections to the grid:

- The government could consider regulations requiring applications for generating plant and grid connections to be submitted jointly (RCEP 2000).
- Government offices in the regions could establish regional forums for the planning of new energy developments (English Nature 2000), with representation from the regions, LPAs, the electricity industry and environmental bodies (Section 10.6.3).
- Strategic levels of planning could be developed as part of "regional spatial strategies", along the lines proposed by the RCEP (2002), which incorporates energy infrastructure (Section 3.3.7).

- The electricity industry could develop methods of strategic environmental assessment in order to assess the overall environmental effects of its current and proposed activities.

3. <u>Information on the effects of HVDT</u>

LPAs' knowledge of the effects of HVDT installations and the possible means of offsetting those effects is generally partial, and inadequate for effective policy formation. Likewise, public understanding of HVDT issues is largely shaped by media-selected topics (Section 10.9.1). LPAs and others would benefit from fuller information, including the analysis of effects undertaken by the electricity industry as presented in their environmental reports and other relevant documents (Section 10.6.1; Figure 2.1).

- The electricity industry could commission independent consultants to produce a publication, or other material, presenting the industry's analysis of HVDT's effects, for dissemination to LPAs.
- The electricity industry could bring greater public attention to its current environmental practice and to technical innovations designed to alleviate the effects of HVDT installations (see the following point).

4. <u>Environmental planning practice and technical innovation</u>

The electricity industry needs to be responsive to the changing perceptions of representative LPAs towards HVDT installations and to increasing public unwillingness to accept the effects that they are seen to have, by seeking improvements in environmental planning practice and remaining open to technical innovations that may assist in this.

- Continued improvements could be made in environmental impact assessment practice (Section 2.9), especially the development of mitigation measures (Marshall 2000).
- Good practice and technical developments could be exchanged between different UK electricity companies, possibly via the Electricity Association.
- Good practice and technical developments could be exchanged internationally, through the International conference on large high voltage electric systems (CIGRE); CIGRE could be encouraged to open a planning group for assistance.
- Selected topics could be investigated and researched to seek means of achieving improved practice. Examples include line compaction (Section 2.11), service corridors (Section 6.12.2), development close to OHLs (Section 10.11), undergrounding (Section 10.6.2) and alternative tower designs.
- The industry could support Ofgem's appeal for environmental considerations to become central to the regulatory framework of the industry (Section 11.3.2), so that improved practice is not immediately precluded on the basis of cost.

5. Development close to HVDT installations

There is a widespread sense that there is a lack of clear government guidance on the planning of development close to HVDT installations; some planning officers and inspectors have called for this to be provided (Section 10.9.5).

- The Office of the Deputy Prime Minster, in consultation with the departments responsible for the draft EMF circular (Section 2.6.2), could produce a guidance note on the planning of development close to HVDT installations; this should deal primarily with EMF exposure levels, but also cover amenity matters, giving guidance on the visual and other effects to be taken into account when planning housing and other development close to HVDT installations.

11.5 HVDT INSTALLATIONS IN THE LOCAL ENVIRONMENT

This book has explored a contentious land-use planning issue and has revealed growing tensions between utility development and environmental interests. The nature of HVDT infrastructure is such that these tensions are pronounced: installations are large scale and industrial in character and extremely dispersed throughout both rural and urban areas. The sense of conflict between HVDT installations and their surroundings has recently come to expression in local authority development plans throughout England and Wales. This is partly due to recent and impending extensions to the grid system resulting from current trends in electricity generation, which have caused growing concern about their effects on the areas where they are located. Conflict has also arisen owing to other development being built in proximity to existing HVDT installations, as a result of economic and housing pressures. But a sensitisation to the effects of installations on their surroundings has been more significant. Many rural areas have been invested with a renewed sense of worth, through the development of environmental strategies in plan-making; residential areas have come to be seen as needing protection from a perceived risk to health and as deserving improved amenity; some derelict urban areas have been attributed regeneration potential. It is this attribution of value to the settings within which HVDT installations are currently found, or may be located, that has heightened the sense of incompatibility between them.

This growing perception of HVDT installations as intrusive elements in the landscape has found expression in a multiplicity of settings, both rural and urban; it has been voiced by a range of different actors, according to the experience of installations in their localities. Policy makers have been able to draw on a range of supportive instruments and precedents, again reflecting local conditions and experience.

The renewed emphasis on development plans since 1991 has provided local authorities with a vehicle to give formal expression to their concerns, and to try to influence the siting of HVDT installations and nearby development. Plan-making procedures themselves have played an important part in the shaping of policy, by providing an arena within which issues have been contested and policy wording has been adapted as a consequence. The preparation of development plans has therefore facilitated the voicing of widespread concerns and has given some opportunity for opposing viewpoints to be presented. There remains the potential, however, for greater understanding of different perspectives on HVDT planning issues to be achieved, especially between local authorities and the electricity industry. It is to this end that more cooperative forms of planning HVDT installations and nearby development should be sought. Ultimately, however, real and mounting public concerns regarding the presence of HVDT installations will be best addressed by continuing improvements in the location, form and design of HVDT and related development. As with other forms of land use that come under increasing challenge because of their serious environmental implications, the aim should be to achieve "legitimate economic and social goals in ways that safeguard, and wherever appropriate, enhance the quality of the environment" (RCEP 2002, p. 38).

References

Adams, D. (1994) *Urban Planning and the Development Process*, UCL Press, London.

Adams, D. and Pawson, G. (1991) *Representation and Influence in Local Planning*, Occasional Paper 31, Department of Planning and Landscape, University of Manchester, Manchester.

Adema, B., Bornebroek, D., Kling, W. and Slofstra, J. (1996) *Experiences with the Realization of New Overhead Transmission Lines in the Netherlands; Technical Options, Legal and Administrative Procedures, Environmental Constraints and Public Acceptance*, Paper 22-207, CIGRE, Paris.

Alexandrov, G., Dikoi, V., Krylov, S., Nikitin, O. and Timashova, L. (1998) *Overhead Line Designing in View of Environmental Constraints: Compact Overhead Lines*, Paper 22/33/36-05, CIGRE, Paris.

Allmendinger, P. and Thomas, H. (eds.) (1998) *Urban Planning and the New British Right*, Routledge, London.

Awad, M., Swidan, M., Said, H. and Moustafa, F. (1998) *The Egyptian Experience for Electric and Magnetic Fields and Radio Noise with Compact and Traditional Towers*, Paper 22/33/36-09, CIGRE, Paris.

Aylesbury Vale District Council. (1991) *Aylesbury Local Plan,* Aylesbury Vale District Council, Aylesbury.

Aylesbury Vale District Council. (1995) *Aylesbury Vale (Rural Areas) Local Plan,* Aylesbury Vale District Council, Aylesbury.

Aylesbury Vale District Council. (1998) *Aylesbury Vale District Local Plan: Deposit Draft,* Aylesbury Vale District Council, Aylesbury.

Baber, P. (1996) Pilloried by the Local Plan Process, *Planning Week*, 4 (30), 9.

Baker, M. (1997) Development Plans: an Agenda for Future Change, *Town and Country Planning*, 66 (9), 242–244.

Barrow, C. (1999) *Environmental Management: Principles and Practice*, Routledge, London.

Bascom, E., Douglass, D. and Thomann, G. (1996) *Hybrid Transmission: Aggressive Use of Underground Cable Sections with Overhead Lines*, Paper 21/22-10, CIGRE, Paris.

Behncke, R., Clark, A., White, H., Opaschaitat, P., Sparrow, L. and Tong, Y. (1998) *500kV Compact Line Design for the Greater Bangkok Area*, Paper 22/33/36-12, CIGRE, Paris.

Bell, S. and McGillivray, D. (2000) *Ball and Bell on Environmental Law: The Law and Policy Relating to the Protection of the Environment*, Blackstone Press, London.

Blowers, A (1980) *The Limits of Power – The Politics of Local Planning Policy*, Pergamon, Oxford.

Bohme, H., Paschen, R., Bertazzi, A., Di Marco, G., Elli, E. and Catencacci, A. (1998) *Overhead Transmission Lines: Design Aimed to Reduce the Permitting Time*, Paper 22/33/36-07, CIGRE, Paris.

Breckland Council. (1999) *Breckland District Local Plan*, Breckland Council, Dereham.

Bridgend County Borough Council. (1997) *Mid Glamorgan Replacement Structure Plan,* Bridgend County Borough Council, Bridgend.

Bridgend County Borough Council. (2000) *Unitary Development Plan Draft Deposit,* Bridgend County Borough Council, Bridgend.

Bridgend County Borough Council. (2001) *Unitary Development Plan Deposit,* Bridgend County Borough Council, Bridgend.

British Waterways. (1991) *British Waterways: the Waterway Environment and Development Plans*, British Waterways, Watford.

Broads Authority. (1997) *Broads Local Plan*, Broads Authority, Norwich.

Bruff, G. and Wood, A. (1995) Sustainable Development in English Metropolitan District Authorities: An Investigation using Unitary Development Plans, *Sustainable Development*, 3, 9–19.

Bruton, M. and Nicholson, D. (1987) *Local Planning in Practice*, Hutchinson, London.

Bryman, A. (1988) *Quantity and Quality in Social Research*, Unwin Hyman, London.

Buckinghamshire County Council. (1996) *Buckinghamshire County Structure Plan 1991–2011*, Buckinghamshire County Council, Aylesbury.

Calderdale Metropolitan Borough Council. (1994) *Calderdale Unitary Development Plan Inspector's Report*, Calderdale Metropolitan Borough Council, Halifax.

Canter, L. (1996) *Environmental Impact Assessment*, 2nd edition, McGraw-Hill, New York.

Carruthers, R. (1987) *Planning Overhead Power Line Routes*, Research Studies Press, Letchworth.

Chadwick, P. and Sienkiewicz, Z. (1998) *Electromagnetic Fields: WHO Draft Document*, WHO, Geneva.

Chris Blandford Associates. (1994) *Gwynedd Renewable Energy Study: The Environmental and Economic Implications*, Chris Blandford Associates, Cardiff.

CIGRE. (2000) *Statutes*, CIGRE, Paris.

CIGRE Joint Working Group. (1996) *Comparison of Overhead Lines and Underground Cables for Electricity Transmission*, Paper 21/22-01, CIGRE, Paris.

CIGRE Joint Working Group. (1998) *Environmental Aspects in Substations*, Paper 23-201, CIGRE, Paris.

Cleveland County Council. (1995) *Cleveland Structure Plan: Alteration no 1*, Cleveland County Council, Cleveland.

Copeland Borough Council. (1995) *The Copeland Local Plan: Inspector's Report*, Copeland Borough Council, Whitehaven.

Council for the Protection of Rural England. (1992) *Campaigners' Guide to Local Plan*s, Council for the Protection of Rural England, London.

Council for the Protection of Rural England. (1995) *Overhead Electricity Lines: Briefing Paper*, Council for the Protection of Rural England, London.

Counsell, D. (1998) Sustainable Development and Structure Plans in England and Wales: A Review of Current Practice, *Journal of Environmental Planning and Management*, 41 (2), 177–194.

Counsell, D. (1999) Sustainable Development and Structure Plans in England and Wales: Operationalizing the Themes and Principles, *Journal of Environmental Planning and Management*, 42 (1), 45–61.

Countryside Agency. (1999) *The State of the Countryside 1999*, Countryside Agency, Cheltenham.

Countryside Commission. (1991) *Fit for the Future: Report of the National Parks Review Panel*, Countryside Commission, Cheltenham.

Countryside Commission. (1994) *Overhead Electricity Lines: Reducing the Impact*, Countryside Commission, Cheltenham.

Countryside Commission, English Heritage and English Nature. (1996) *Conservation Issues in Local Plans*, English Heritage, London.

Crouch, M. (1995) Wider Review Needed to get Plans Back on Track, *Planning*, 17 February 1995, 1106.

Cullingworth, J. and Nadin, V. (2002) *Town and Country Planning in the UK*, 13th edition, Routledge, London.

D'Adjello, L., Cauzillo, B., Nicolini, P., Paris, L., Biondi, G., Guastadisegni, E. and Piendibene, G. (1996) *Quantitative Evaluation of the Visual Impact of Overhead Lines*, Paper 22-210, CIGRE, Paris.

Davies, H. (1999) The Planning System and the Development Plan, in Cullingworth, B. (ed.) *British Planning: 50 Years of Urban and Regional Policy*, 45–61, Athlone Press, London.

Davoudi, S., Hull, A. and Healey, P. (1996) Environmental Concerns and Economic Imperatives in Strategic Plan Making, *Town Planning Review*, 67 (4), 421–435.

Deasley, N. and Hamilton, J. (1995) Implementability – Has Your Plan Got What it Takes? *Planning*, 24 February 2005, 1107.

Denzin, N. and Lincoln, Y. (1998) Introduction: Entering the Field of Qualitative Research, in Denzin, N. and Lincoln, Y. (eds.), *Collecting and Interpreting Qualitative Materials*, 1–34, Sage Publications, London.

Department of the Environment. (1974) *Report of the National Park Policies Review Committee*, HMSO, London.

Department of the Environment. (1985) *Development and Employment: Circular 14/85*, HMSO, London.

Department of the Environment. (1989) *Regional Planning Guidance: Strategic Guidance for South Yorkshire: RPG 5*, HMSO, London.

Department of the Environment. (1990a) *Electricity Generating Stations and Overhead Lines: Circular 14/90*, HMSO, London.

Department of the Environment. (1990b) *This Common Inheritance*, Cm 1200, HMSO, London.

Department of the Environment. (1992a) *Planning Policy Guidance Note 12: Development Plans and Regional Planning Guidance*, HMSO, London.

Department of the Environment. (1992b) *Planning Policy Guidance Note 1: General Policy and Principles*, HMSO, London.

Department of the Environment. (1992c) *Planning Policy Guidance Note 7: The Countryside and the Rural Economy*, HMSO, London.

Department of the Environment. (1993a) *Regional Planning Guidance for the Northern Region: RPG 7*, HMSO, London.

Department of the Environment. (1993b) *Planning Policy Guidance Note 22: Renewable Energy*, HMSO, London.

Department of the Environment. (1994) *Planning Policy Guidance Note 15: Planning and the Historic Environment*, HMSO, London.

Department of the Environment. (1995) *The Thames Gateway Planning Framework: RPG9a*, HMSO, London.

Department of the Environment. (1997a) *Planning Policy Guidance Note 1 (Revised): General Policy and Principles*, The Stationery Office, London.

Department of the Environment. (1997b) *Planning Policy Guidance Note 7 (Revised): The Countryside – Environmental Quality and Economic and Social Development*, The Stationery Office, London.

Department of the Environment. (1997c) *Planning Policy Guidance Note 23: Planning and Pollution Control*, The Stationery Office, London

Department of the Environment, Transport and the Regions. (1998) *Planning for Sustainable Development: Towards Better Practice*, DETR, London.

Department of the Environment, Transport and the Regions. (1999a) *Planning Policy Guidance Note 12: Development Plans*, The Stationery Office, London.

Department of the Environment, Transport and the Regions. (1999b) *Projections of Households in England to 2021: 1996-Based Estimates of the Numbers of Households for Regions*, DETR, London.

Department of the Environment, Transport and the Regions. (1999c) *A Better Quality of Life: A Strategy for Sustainable Development for the UK*, Cm 4345, The Stationery Office, London.

Department of the Environment, Transport and the Regions and Department of Health. (1998) *Land Use Planning and Electromagnetic Fields (EMFs): Consultation Draft*, DETR and DoH, London.

Department of Trade and Industry. (1998) *Conclusions of the Review of Energy Sources for Power Generation and Government Response to Fourth and Fifth Reports of the Trade and Industry Committee*, Cm 4071, The Stationery Office, London.

Department of Trade and Industry. (2000) *Guidance on the Electricity Works (Environmental Impacts Assessment) (England and Wales) Regulations 2000*, The Stationery Office, London.

Department of Trade and Industry. (2001) *New and Renewable Energy: Prospects for the 21st Century: Conclusions in Response to the Public Consultation*, DTI, London.

Department of Trade and Industry. (2002) *Future Offshore: A Strategic Framework for the Offshore Wind Energy*, DTI, London.

Department of Trade and Industry. (2003) *Our Energy Future – Creating a Low Carbon Economy*, Cm 5761, Stationery Office, London.

Department for Transport, Local Government and the Regions. (2001a) *Planning Green Paper: Planning – Delivering a Fundamental Change*, DTLR, London.

Department for Transport, Local Government and the Regions. (2001b) *Planning Policy Guidance Note 8 (Revised): Telecommunications*, The Stationery Office, London.

Doyen-Viguier, C., Gaudry, M. and Laurent, G. (1996) *Digital Simulation for Better Integration of Overhead Transmission Lines in the Environment*, Paper 22-201, CIGRE, Paris.

Edwards, M. and McCafferty, A. (1992) 1991: A Time to Reflect, *Estates Gazette* 9223, 13 June, 103–106.

Egger, H. and Lugschitz, H. (1998) *Experience with a 380kV 'Camouflage Line'*, Paper 22/33/36-04, CIGRE, Paris.

Electricity Association. (1988) *Technical Specification 43-8: Overhead Line Clearances*, Electricity Association, London.

Energy Technology Support Unit. (1997) *Eastern Region Renewable Energy Planning Study: Final Report*, ETSU, Didcot.

English Nature. (2000) *Position Statement on Electricity Generation, Transmission and Use*, English Nature, Peterborough.

Engsbro, L., Lorensen, L., Dam-Anderson, M. and Kaad Jensen, J. (1996) *Modernization of the High Voltage Transmission System in Denmark*, Paper 21/22-03, CIGRE, Paris.

Ernst, J. (1994) *Whose Utility? The Social Impact of Public Utility Privatization and Regulation in Britain*, Open University Press, Buckingham.

Fair, J. (1994) Pressure Groups Push Up the Cost of UDPs, *Planning Week*, 2 (14), 8.

Fitzgerald, C., Bailey, R. and Flanagan, P. (1996) *Community Consultation and Environmental Assessment for High voltage Transmission Line Projects in New South Wales*, Paper 22-202, CIGRE, Paris.

Flick, U. (1998) *An Introduction to Qualitative Research*, Sage, London.

Friends of the Earth. (1994) *Planning for the Planet: Sustainable Development Policies for Local and Strategic Plans*, Friends of the Earth, London.

Furby, L., Slovic, P., Fischhoff, B. and Gregory, R. (1988) Public Perceptions of Electric Power Transmission Lines, *Journal of Environmental Psychology*, 8, 19–43.

Gatenby, I. and Williams, C. (1996) Interpreting Planning Law, in Tewdwr-Jones, M. (ed.) *British Planning Policy in Transition; Planning in the 1990s*, 137–153, UCL Press, London.

Gibbs, D., Longhurst, J. and Braithwaite, C. (1998) `Struggling with Sustainability': Weak and Strong Interpretations of Sustainable Development within Local Authority Policy, *Environment and Planning A*, 30, 1351–1365.

Goulty, G. (1990) *Visual Amenity Aspects of High Voltage Transmission*, Research Studies Press, Taunton.

Government Office for Yorkshire and Humberside. (1996) *Regional Planning Guidance for Yorkshire and Humberside: RPG 12*, HMSO, London.

Great Yarmouth Borough Council. (2000) *Great Yarmouth Borough-wide Local Plan*, Great Yarmouth Borough Council, Great Yarmouth.

Hales, R. (2000) Land Use Development Planning and the Notion of Sustainable Development: Exploring Constraint and Facilitation within the English Planning System, *Journal of Environmental Planning and Management*, 43(1), 99–121.

Hall, P. (2002) Pull Lines Down and Watch New Schemes Rise, *Regeneration and Renewal*, 5 April 2002, 10.

Healey, P. (1983) *Local Plans in British Land Use Planning*, Pergamon, Oxford.

Healey, P. (1994) Development Plans: New Approaches to Making Frameworks for Land Use Regulation, *European Planning Studies*, 2, 39–57.

Healey, P. (1997a) An Institutionalist Approach to Spatial Planning, in Healey, P., Khakee, A., Motte, A. and Needham, B. (eds.) *Making Strategic Spatial Plans: Innovation in Europe*, 21–36, UCL Press, London.

Healey, P. (1997b) *Collaborative Planning*, Macmillan, Basingstoke.

Healey, P., McNamara, P., Elson, M. and Doak, A. (1988) *Land Use Planning and the Mediation of Urban Change*, Cambridge University Press, Cambridge.

Healey, P. and Shaw, T. (1993) Planners, Plans and Sustainable Development, *Regional Studies*, 27, 769–776.

Healey, P. and Shaw, T. (1994) Changing Meanings of `Environment' in the British Planning System, *Transactions of the Institute of British Geographers*, 19, 425–438.

Henning, G., Eriksson, A. and Jonsson, U. (1996) *Compacted Overhead Lines with Low Magnetic Fields*, Paper 22-204, CIGRE, Paris.

Her Majesty's Government. (1990) *This Common Inheritance: Britain's Environmental Strategy*, Cm 1200, HMSO, London.

Her Majesty's Government. (1994) *The UK Strategy for Sustainable Development*, Cm 2426, HMSO, London.

Hill, M. (1997) *Studying the Policy Process*, Prentice Hall, Harlow.

Hughes, D. (1997) *Environmental Law*, Butterworths, London.

Hull, A. and Vigar, G. (1998a) The Changing Role of the Development Plan in Managing Spatial Change, *Environment and Planning C*, 16 (4), 379–394.

Hull, A. and Vigar, G. (1998b) Development Plans: Coping with Re-regulation in the 1990s, in Allmendinger, P. and Thomas, H. (eds.) *Urban Planning and the New British Right*, 114–136, Routledge, London.

Hunt, D. and Johnson, C. (1995) *Environmental Management Systems*, McGraw-Hill, London.

Independent Expert Group on Mobile Phones. (2000) *Mobile Phones and Health*, NRPB, Didcot.

Institution of Electrical Engineers. (2001) *Factfile: Electromagnetic Fields and Health*, IEE, London.

International Energy Agency. (2001) *Key World Energy Statistics from the IEA*, International Energy Agency, Paris.

International Organisation for Standardisation. (1996) *IS0 14001:1996 Environmental Management Systems: Specification with Guidance for Use*, British Standards Institution, London.

Jans, J. (1995) *European Environmental Law*, Kluwer Law International, London.

John, P. (1998) *Analysing Public Policy*, Pinter, London.

Kikuchi, T., Oba, W., Kojima, Y., Asano, Y. and Matsui, S. (1998) *Compact Transmission Lines for Increasing Voltage while Keeping Existing Equipment Intact*, Paper 22/33/36-03, CIGRE, Paris.

Kim, J., Cho, S., Lee, D. and Shim, E. (1996) *Route Selection and Design of 765kV Transmission Line Considering Environmental Constraints and Public Hearing*, Paper 22-208, CIGRE, Paris.

Landscape Institute and Institute of Environmental Management and Assessment. (2002) *Guidelines for Landscape and Visual Impact Assessment*, Spon Press, London.

Langbaurgh on Tees Borough Council. (1991) *Planning Policy Guidance: Siting of Overhead Power Lines*, Langbaurgh on Tees Borough Council, South Bank.

Langbaurgh on Tees Borough Council. (1995) *Langbaurgh on Tees Local Plan: Deposit Document*, Langbaurgh on Tees Borough Council, South Bank.

Lincoln City Council. (1997) *The City of Lincoln Local Plan: Inspector's Report of Objections*, Lincoln City Council, Lincoln.

Local Government Association. (1999) *LGA Response to Draft Advice to Local Planning Authorities on Land-Use Planning and Development Giving Rise to Electromagnetic Fields*, LGA.

Lomas, C., Jackson, R. and Speed, G. (1996) *Integrating Overhead Lines into an Environmentally Sensitive World*, Paper 22-206, CIGRE, Paris.

London Borough of Merton. (2001) *London Borough of Merton Council UDP: Public Local Inquiry 2001: Response Proof*, London Borough of Merton, London.

London Borough of Newham. (1980) *Beckton Local Plan*, London Borough of Newham, London.

London Borough of Newham. (1985a) *Beckton Local Plan (Draft)*, London Borough of Newham, London.

London Borough of Newham. (1985b) *South Docklands Local Plan (Draft)*, London Borough of Newham, London.

Loudon, D., Halsam, K., Jonsson, U., Karlsson, D., Stensrom, L. and Lundquist, J. (1998) *A Compact 420kV Line Utilising Line Surge Arresters for Areas with Low Isokeraunic Levels*, Paper 22/33/36-08, CIGRE, Paris.

Luain, C. and Corcoran, A. (1996) *Transmission Line Route Selection Procedures for Minimal Environmental Impact in Ireland*, Paper 22-203, CIGRE, Paris.

MacEwan, A. and MacEwan, M. (1982) *National Parks: Conservation or Cosmetics?* George Allen and Unwin, London.

MacGregor, B. and Ross, A. (1995) Master or Servant? The Changing Role of the Development Plan in the British Planning System, *Town Planning Review*, 66 (1), 41–60.

Marshall R. (2000) The Application of Mitigation and its Resolution within EIA: the Scottish Power Perspective, *20th IAIA Conference Proceedings*, IAIA, Hong Kong.

Mendip District Council. (1998) *Mendip District Local Plan: Deposit Draft*, Mendip District Council, Shepton Mallett.

Messing, M., Friesema, H. and Morell, D. (1979) *Centralized Power: the Politics of Scale in Electricity Generation*, Oelgeschlager, Gunn and Hain, Massachusetts, Cambridge, MA.

Miles, M. and Huberman, A. (1994) *Qualitative Data Analysis: an Expanded Sourcebook*, Sage Publications, London.

Miller, C. and Fricker, C. (1993) Planning and Hazard, *Progress in Planning*, 40 (3), 167–260.

Millichap, D. (1995) Law, Myth and Community: a Reinterpretation of Planning's Justification and Rationale, *Planning Perspectives*, 10, 279–293.

Milne, R. (1997) Radical Rethink in Prospect as Survey Backs Plan Lead, *Planning*, 17 January 1997 (Weekly journal), 1201.

Moore, V. (2000) *A Practical Approach to Planning Law*, Blackstone Press, London.

Myerson, G. and Rydin, Y. (1996) Sustainable Development: The Implications of the Global Debate for Land Use Planning, in Buckingham-Hatfield, S. and Evans, B. (eds.) *Environmental Planning and Sustainability*, 19–34, John Wiley, Chichester.

National Grid. (unpublished) *The National Grid Company and New High Voltage Transmission Lines: Guidelines for Line Routeing (the Holford Rules) and Undergrounding* (internal document), The National Grid Group, Coventry.

National Grid. (unpublished) *NGC Substations and the Environment: Guidelines on Siting and Design* (internal document), The National Grid Group, Coventry.

National Grid. (unpublished) *Undergrounding of High Voltage Cables: NGC's Policy and Guidelines* (internal document), The National Grid Group, Coventry.

National Grid. (unpublished) *Overhead or Underground? The National Grid Company Approach*, The National Grid Group, Coventry.

National Grid. (1998) *Group Environmental Performance Report 1997/1998*, The National Grid Group, Coventry.

National Grid. (1999a) *1999 Seven Year Statement: For the Years 1999/00 to 2005/2006*, The National Grid Group, Coventry.

National Grid. (1999b) *The Management of our Environment: Environmental Performance Report 1998/99,* The National Grid Group, Coventry.

National Grid. (1999c) *Extracts from Consents and Wayleaves Manual* (internal document), The National Grid Group, Coventry.

National Grid. (2000) *National Grid: UK Operations Environmental Assessment Guidance Manual* (internal document), The National Grid Group, Coventry.

National Grid. (2001a) *The National Grid Company Electricity Act 1989 Schedule 9 Statement: Duty of Preservation and Amenity*, The National Grid Group, Coventry.

National Grid. (2001b) *Development Near Lines: Planning and Amenity Aspects of High Voltage Transmission Lines and Substations*, The National Grid Group, Coventry.

National Grid. (2001c) *National Grid and Distributed Generation: Facilitating the Future*, The National Grid Group, Coventry.

National Radiological Protection Board. (2001) *Report of an Advisory Group on Non-ionising Radiation: ELF Electromagnetic Fields and the Risk of Cancer*, Doc. NRPB 12 (1), NRPB, Didcot.

Netherwood, A. (1996) Environmental Management Systems, in Welford, R. (ed.) *Corporate Environmental Management: Systems and Strategies*, 35–58, Earthscan, London.

Newby, H. (1990) Ecology, Amenity and Society: Social Science and Environmental Change, *Town Planning Review*, 61 (1), 3–13.

Newham Council. (1993) *Newham's Unitary Development Plan Deposit Draft*, Newham Council, London.

Newham Council. (1995) *Report on Objections to the Deposit Draft of the Newham Unitary Development Plan*, Newham Council, London.

Newham Council. (1997) *Newham's Unitary Development Plan Adopted Version*, Newham Council, London.

Newham Council. (2001a) *The Newham Unitary Development Plan*, Newham Council, London.

Newham Council. (2001b) *Lower Lea Valley, Stratford to Thameside: Planning Framework*, Newham Council, London.

Newham Council. (2001c) *Canning Town Area Action Plan*, Newham Council, London.

Norfolk County Council. (1996) *Reports to the Planning Sub-Committee*, Norfolk County Council, Norwich.

Norfolk County Council. (1997) *Draft Norfolk Structure Plan*, Norfolk County Council, Norwich.

Norfolk County Council. (1998) *Norfolk Structure Plan Review: Examination in Public, Report of the Panel*, Norfolk County Council, Norwich.

Norfolk County Council. (1999) *Norfolk Structure Plan*, Norfolk County Council, Norwich.

Northern Electric and Gas. (1998) *Annual Environmental Performance Report 1998*, Northern Electric and Gas Ltd, Newcastle upon Tyne.

North Norfolk District Council. (1998) *North Norfolk Local Plan*, North Norfolk District Council, Cromer.

Northumberland County Council. (1996) *Northumberland County Structure Plan*, Northumberland County Council, Morpeth.

Ofgem. (2001) *Environmental Action Plan*, Ofgem, London.

Ogwr Borough Council. (1995) *Ogwr Borough Local Plan*, Ogwr Borough Council, Bridgend.

Ogwr Borough Council. (1996) *Development Brief for Broadlands, South West Bridgend*, Ogwr Borough Council, Bridgend.

Ojala, Y., Leskinen, T., Lahtinen, M. and Hinkkuri, A. (1998) *110kV Overhead Transmission Line with Covered Conductors*, Paper 22/33/36-10, CIGRE, Paris.

O'Riordan, T. (1995) *Environmental Science for Environmental Management*, Longman, Harlow.

O'Riordan, T. and Jordan, A. (1995) The Precautionary Principle in Contemporary Environmental Politics, *Environmental Values*, 4, 191–212.

Owens, S. (1994) Land, Limits and Sustainability: a Conceptual Framework and Some Dilemmas for the Planning System, *Transactions of the Institute of British Geographers*, 19, 439–456.

Owens, S. (1997a) 'Giants in the Path': Planning, Sustainability and Environmental Values, *Town Planning Review*, 68 (3), 293–304.

Owens, S. (1997b) Negotiated Environments? Needs, Demands, and Values in the Age of Sustainability, *Environment and Planning A*, 29, 571–580.

Owens, S. and Cowell, R. (2002) *Land and Limits: Interpreting Sustainability in the Planning Process*, Routledge, London.

Parsons, W. (1995) *Public Policy: An Introduction to the Theory and Practice of Policy Analysis*, Edward Elgar, Aldershot.

Patterson, W. (1999) *Transforming Electricity: The Coming Generation of Change*, Earthscan, London.

Performance and Innovation Unit. (2002) *The Energy Review*, PIU, London.

Planning Inspectorate. (1992) *Applications by the National Grid Company for Grant of Necessary Wayleaves for the East London 400kV Line*, Department of the Environment, London.

Priestley, T. and Evans, G. (1996) Resident Perceptions of a Nearby Electric Transmission Line, *Journal of Environmental Psychology*, 16, 65–74.

Redcar & Cleveland Borough Council. (1999) *Redcar & Cleveland Local Plan*, Redcar & Cleveland Borough Council, South Bank.

Robson, C. (2002) *Real World Research: A Resource for Social Scientists and Practitioner-Researchers*, Blackwell, Oxford.

Rogier, J., Goosens, L., Mazoin, M., Robberechts, W., De Clerck, E. and Jadot, A. (1996) *Visualization of Overhead Line Projects*, Paper 22-209, CIGRE, Paris.

Rotherham Metropolitan Borough Council. (1991) *Rotherham Metropolitan Borough Council Unitary Development Plan: A Discussion Document about the Regeneration of Rotherham into the 21st Century*, Rotherham Metropolitan Borough Council, Rotherham.

Rotherham Metropolitan Borough Council. (1994) *Rotherham Metropolitan Borough Council Unitary Development Plan: Consultation Draft*, Rotherham Metropolitan Borough Council, Rotherham.

Rotherham Metropolitan Borough Council. (1995) *Rotherham Metropolitan Borough Council Unitary Development Plan: Deposit*, Rotherham Metropolitan Borough Council, Rotherham.

Rotherham Metropolitan Borough Council. (1996) *Proposed Changes to the Deposit UDP*, Rotherham Metropolitan Borough Council, Rotherham.

Rotherham Metropolitan Borough Council. (1999) *Rotherham Metropolitan Borough Council Unitary Development Plan: Adopted Version*, Rotherham Metropolitan Borough Council, Rotherham.

Royal Commission on Environmental Pollution. (2000) *Energy – the Changing Climate; Twenty-Second Report*, Cm 4749, The Stationery Office, London.

Royal Commission on Environmental Pollution. (2002) *Twenty-Third Report: Environmental Planning*, Cm 5459, The Stationery Office, London.

Rydin, Y. (1998) *Urban and Environmental Planning in the UK*, Macmillan, Basingstoke.

Rydin, Y. (1999) Public Participation in Planning, in Cullingworth, B. (ed.) *British Planning: 50 Years of Urban and Regional Policy*, 184–197, Athlone Press, London.

Scott, P., Miller, C. and Wood, C. (1998) Planning and Pollution: an Unusual Perspective on Central – Local Relations, *Environment and Planning C*, 16, 529–542.

Scottish and Southern Energy. (1999) *Environmental Report 1999*, Scottish and Southern Energy, Perth.

Sheail, J. (1992) The Amenity Clause: an Insight into Half a Century of Environmental Protection in the United Kingdom, *Transactions of the Institute of British Geographers*, 17, 152–165.

Sheate, W. (1995) Electricity Generation and Transmission: a Case Study of Problematic EIA Implementation in the UK, *Environmental Policy and Practice*, 5 (1), 17–25.

Snowdonia Joint Advisory Committee. (1971) *Eighteenth Annual Report 1970/197*, Snowdonia National Park, Dolgellau.

Snowdonia National Park Authority. (1977) *Snowdonia National Park Plan*, Snowdonia National Park Authority, Penrhyndeudraeth.

Snowdonia National Park Authority. (1986) *Snowdonia National Park Plan First Review*, Snowdonia National Park Authority, Penrhyndeudraeth.

Snowdonia National Park Authority. (1991) *Snowdonia National Park Conservation Map*, Snowdonia National Park Authority, Penrhyndeudraeth.

Snowdonia National Park Authority. (1993) *Energy*, Snowdonia National Park Authority, Penrhyndeudraeth.

Snowdonia National Park Authority. (1997a) *Response to Representations on Consultation Draft*, Snowdonia National Park Authority, Penrhyndeudraeth.

Snowdonia National Park Authority. (1997b) *Eryri Local Plan: Deposit Version*, Snowdonia National Park Authority, Penrhyndeudraeth.

Snowdonia National Park Authority. (1999) *Eryri Local Plan*, Snowdonia National Park Authority, Penrhyndeudraeth.

Somerset County Council. (1997) *Somerset Structure Plan Review: Deposit Plan*, Somerset County Council, Taunton.

South Norfolk Council. (1997) *South Norfolk Local Plan*, South Norfolk Council, Cromer.

South Wales Electricity. (unpublished) *Statutory Requirements – Our Schedule 9 Statement*, SWE, Cardiff.

South Western Electricity. (unpublished) *Schedule 9 Statement*, SWEB, Bristol.

Stake, R. (1998) Case Studies, in Denzin, N. and Lincoln, Y. (eds.) *Strategies of Qualitative Inquiry*, 86–109, Sage Publications, London.

Surrey, J. (ed.) (1996) *The British Electricity Experiment: Privatisation: The Record, the Issues, the Lessons*, Earthscan, London.

Swansea City and County. (1998) *Swansea Local Plan: Inspector's Report*, Swansea City and County, Swansea.

Swindon Borough Council (SBC). (1999) *Swindon Borough Local Plan*, Swindon Borough Council, Swindon.

Tees Valley Joint Strategy Unit. (2000) *Tees Valley Structure Plan*, Tees Valley Joint Strategy Unit, Middlesbrough.

Terence O'Rourke. (1997) *Eastern Region Renewable Energy Planning Study: Final Report*, Terence O'Rourke, Bournemouth.

Tewdwr-Jones, M. (1994a) The Development Plan in Policy Implementation, *Environment and Planning C*, 12 (2), 145–164.

Tewdwr-Jones, M. (1994b) Policy Implications of the 'Plan-Led' Planning System, *Journal of Planning and Environment Law*, 5, 584–593.

Tewdwr-Jones, M. (1996) Land Use Planning Policy after Thatcher, in Tewdwr-Jones, M. (ed.) *British Planning Policy in Transition; Planning in the 1990s*, 1–13, UCL Press, London.

Thamesdown Borough Council. (1987) *Notice of Meeting of the Planning Committee 2nd July 1987*, Thamesdown Borough Council, Swindon.

Thamesdown Borough Council. (1993) *Notice of Meeting of the Planning Committee 15th March 1993*, Thamesdown Borough Council, Swindon.

Thamesdown Borough Council. (1992) *Supplementary Planning Guidance: A Guide to Good Residential Development*, Thamesdown Borough Council, Swindon.

Thamesdown Borough Council. (1994) *Thamesdown Local Plan: Deposit Draft*, Thamesdown Borough Council, Swindon.

Thamesdown Borough Council. (1995a) *Thamesdown Local Plan: Deposit Draft 1994: List of Proposed Changes*, Thamesdown Borough Council, Swindon.

Thamesdown Borough Council. (1995b) *Thamesdown Local Plan Inquiry Inspector's Report*, Thamesdown Borough Council, Swindon.

Thamesdown Environmental Forum. (1996) *Sustainable Housing Development in Swindon: A Best Practice Guide for Developers*, Thamesdown Environmental Forum, Swindon.

Thomas, S. (1996) The Privatization of the Electricity Supply Industry, in Surrey, J. (ed.) *The British Electricity Experiment: Privatisation: The Record, the Issues, the Lessons*, 40–63, Earthscan, London.

Toshirou, T., Noriaki, F., Yasuo, M. and Masamichi, Y. (1996) *Scenery Preservation Measures and Evaluation Methods for Transmission Lines*, Paper 22-205, CIGRE, Paris.

Tynedale Council. (1994) *Tynedale District Local Plan: Consultation Draft*, Tynedale Council, Hexham.

Tynedale Council. (1996) *Tynedale District Local Plan: Deposit Draft*, Tynedale Council, Hexham.

Tynedale Council. (1998a) *Public Local Inquiry Topic Paper: Area of High Landscape Value*, Tynedale Council, Hexham.

Tynedale Council. (1998b) *Public Local Inquiry Proof of Evidence: Overhead Electricity Power Lines*, Tynedale Council, Hexham.

Tynedale Council. (2000) *Tynedale District Local Plan*, Tynedale Council, Hexham.

UK Centre for Economic and Environmental Development. (1991) Undergrounding Overhead Power Lines: Final Report to the Countryside Commission, UK CEED, Cambridge.

Welsh Office. (1999) *Planning Guidance (Wales): Planning Policy, First Revision*, Welsh Office, Cardiff.

Wilson, E. (1998) Planning and Environmentalism in the 1990s, in Allmendinger, P. and Thomas, H. (eds.) *Urban Planning and the New British Right*, 21–52, Routledge, London.

Wood, C. (2003) *Environmental Impact Assessment: A Comparative Review*, 2nd edition, Longman, Harlow.

Wood, M. (2001) The Vexed Problem of Telephone Mast Fears, *Planning*, 3 August 20.

World Bank. (2000) *Entering the 21st Century: World Development Report 1999/2000*, Oxford University Press, New York.

World Commission on Environment and Development. (1987) *Our Common Future*, Oxford University Press, Oxford.

World Health Organisation. (2000) *Electromagnetic Fields and Public Health: Cautionary Policies*, WHO, Geneva.

Yorkshire Electricity. (1999) *Corporate Environmental Policy and Electricity Act Schedule 9 Statement*, Yorkshire Electricity Group, Leeds.

Yin, R. (1994) *Case Study Research: Design and Methods*, Sage, London.

Statutes

UK Legislation

Electricity Act, 1957

Electricity Act, 1989

Environment Act, 1995

Local Government Act, 1972

National Parks and Access to the Countryside Act, 1949

Planning and Compensation Act, 1991

Planning and Compulsory Purchase Act, 2004

Town and Country Planning Act, 1947

Town and Country Planning Act, 1990

European Legislation

EC Directive 85/337 (Assessment of the Effects of Certain Public and Private Projects on the Environment)

EC Directive 97/11 (Assessment of the Effects of Certain Public and Private Projects on the Environment)

Directive 2001/42/EC (Assessment of the Effects of Certain Plans and Programmes on the Environment)

Statutory Instruments

Electricity Supply Regulations, 1988 (SI 1998 No. 1057)

Electricity Works (Environmental Impact Assessment) (England and Wales) Regulations, 2000 (SI 2000 No. 1927)

NC former Planning Officer 2001, "Communicated by interview".

NC Planning Officer 2001, "Communicated by interview".

NC Elected Member 2001, "Communicated by interview".

NC Regeneration Officer 2001, "Communicated by interview".

NCC Planning Officer 2001, "Communicated by interview".

NCC Lanscape Architect 2001, "Communicated by interview".

NCC Elected Member 2001, "Communicated by interview".

Planning (Hazardous Substances) Regulations, 1992 (SI 1992 No. 656)

RMBC Elected Member 2001, "Communicated by interview".

RMBC Planning Officer A 2000, "Communicated by interview".

RMBC Planning Officer A 2000, "Communicated by interview".

RMBC Planning Officer A and B 2001, "Communicated by interview".

SNPA Planning Officer 2001, "Communicated by interview".

Swindon FoE A and B 2001, "Communicated by interview".

SBC Planning Officer A and B 2001, "Communicated by interview".

SBC Planning Officer C and B 2001, "Communicated by interview".

Town and Country Planning (General Permitted Development) Order, 1995 (SI 1995 No. 418)

TC Planning Officers A and B 2001, "Communicated by interview".

TC Elected Member 2001, "Communicated by interview".

Appendix

AFFECTED ZONE	OPERATIONAL ASPECT	ENVIRONMENTAL EFFECT	NATIONAL GRID (1998)	NATIONAL GRID (1999B)	SCOTTISH AND SOUTHERN ENERGY (1999)	NORTHERN ELECTRIC AND GAS (1998)
AIR AND CLIMATE	Sulphur hexafluoride: used as an insulant at substations	Leaks from equipment: a greenhouse gas	Increasing usage; substantial and increasing losses to the atmosphere		Losses to atmosphere	Measures to minimise leakage
	Site-based energy use	Production of greenhouse gases		Estimates of emissions; decreasing total		
	Vehicle use	Production of greenhouse gases		Not a significant issue		
	Halon: within fire-fighting equipment	Release during usage: an ozone-depleting substance	Incident causing release			
GROUND AND WATER	Oil: used as an insulant and cool ant at substations	Risk of leakage and spillage: ground and water pollution	Programme of pollution prevention by improving oil containment by bunds, interceptors, etc. Possibility of existing ground contamination in old substations; remedial measures as necessary		Bunding of substations	Pollution prevention measures, including bunds and spill kits
	Polychlorinated biphenyls (PCBs): introduced into insulating oils through recycling	Risk of oil leakage and spillage: ground and water contamination, take-up by living organisms	Programme of equipment analysis and of decontamination and removal of equipment, exceeding requirements of proposed legislation			Survey of equipment
	Cable oil: used as an insulant in underground cables	Risk of leakage: ground and water pollution	Estimated leakages; reduced leakages being achieved through effective monitoring and maintenance; research into oil leak detection and remediation	Major leak incident	Leakage rate	Monitoring procedures; possible alternatives to oil

Environmental effects of HVDT described in a sample of corporate environment reports (see Section 2.8)

AFFEC-TED ZONE	OPERATIONAL ASPECT	ENVIRONMENTAL EFFECT	NATIONAL GRID (1998)	NATIONAL GRID (1999B)	SCOTTISH AND SOUTHERN ENERGY (1999)	NORTHERN ELECTRIC AND GAS (1998)
LAND	Pesticides: used to control vegetation below overhead lines and at substations	Risk of harm to living organisms			Estimates of pesticides used	
	Construction and maintenance work	Disturbance to land	Ensuring minimum disturbance to grantors' land; special concern for areas of conservation value and National Parks		Waste from trench spoil	
	Management of non-operational land	Possible harm to land	Contractors set standards for maintenance			
LANDSCAPE	Overhead lines and towers	Visual intrusion on the landscape	Research into visually improved tower designs			Care taken in designated areas; routeing studies, substation design, possible use of overhead cables, undergrounding in sensitive areas
HABITATS AND WILD LIFE	Overhead lines	Bird collisions with conductors: bird deaths	Research into methods of preventing birds coming into contact with overhead lines		Number of incidents; use of bird diverters, especially in a sensitive area	
	Overhead line corridors	Cutting back of shrubs and trees: visual effect			Softening edges by allowing shrubs to grow within the corridor	
	Construction work	Disturbance to sensitive and designated areas			Consultation with conservation bodies to avoid harm	

Environmental effects of HVDT described in a sample of corporate environment reports (see Section 2.8) (continued)

HERITAGE	Construction work	Disturbance to archaeological sites and artefacts			Consultation with archaeological bodies to avoid harm	
HUMAN HEALTH AND NUISANCE	Electric and magnetic fields (EMFs): emitted by conductors	Perceived risk to human health, especially to nearby residents	Recognition of public concern regarding health risks, although research has not established any link, as supported by the National Radiological Protection Board (NRPB); NG's EMF unit liases with the public		NRPB advice followed; practice of responding to public inquiries	Research findings and NRPB advice followed; advice offered to customers
	Noise generated by substation equipment and overhead lines	Nuisance to nearby residents etc.	Seek to ensure noise is kept to a practicable minimum; research on methods of reducing noise		Record of complaints; innovative solution tried	
RESOURCES	Plant operations	Water consumption	Aim to reduce water usage			
	Plant operations	Energy use in substations, offices etc.	Aim to reduce energy consumption by the company; opportunities identified			Aim to reduce energy consumption
	Electricity transmission	Energy losses in transmission	Ongoing reduction in transmission losses through equipment improvements			
	Operations	Waste production	Waste management strategy being developed to minimise waste and increase reuse and recycling		Estimates of waste produced, reused, recycled and disposed of	Waste management strategy being developed

Environmental effects of HVDT described in a sample of corporate environment reports (see Section 2.8) (continued)

	ELECTRICITY TRANSMISSION PROJECTS							
Project reference	Port Talbot	Thorp Arch	Winning-ton	S. Humberside	Isle of Grain	Nunthorpe	Blyth	Newby
Project type	275 kV OHL	275 kV OHL and substation	400 kV OHL	400 kV OHL	400 kV OHL	400 kV underground cable	Substation	Sealing-end compound
County	W. Glamor-gan	N. York-shire	Cheshire	Humberside	Kent	N. Yorkshire	Northum-berland	N. Yorkshire
Date	1997	1994	1995	1992	1998	1999	1997	1998
ENVIRONMENTAL EFFECTS								
Landscape and Visual								
Landscape and visual	•	•	•	•	•	•	•	•
Ground and Water								
Ground					•	•		•
Water quality					•	•		
Drainage						•		•
Habitats and Wildlife								
Flora and fauna	•	•	•	•	•	•	•	•
Birdstrike				•	•			
Cultural Heritage								
Archaeology		•	•	•	•	•	•	•
History and listed buildings	•		•	•	•			
Human Health and Nuisance								
Electric and magnetic fields	•	•	•	•	•	•		•
Noise	•	•	•	•	•	•		•

Environmental effects considered in National Grid environmental statements and reports (see Section 2.9)

	ELECTRICITY TRANSMISSION PROJECTS							
Project reference	Port Talbot	Thorp Arch	Winning-ton	S. Humberside	Isle of Grain	Nunthorpe	Blyth	Newby
Project type	275 kV OHL	275 kV OHL and substation	400 kV OHL	400 kV OHL	400 kV OHL	400 kV underground cable	Substation	Sealing-end compound
County	W. Glamor-gan	N. York-shire	Cheshire	Humberside	Kent	N. Yorkshire	Northumbe rland	N. Yorkshire
Date	1997	1994	1995	1992	1998	1999	1997	1998
ENVIRONMENTAL EFFECTS								
Electromagnetic compatibility	•		•	•	•	•		•
Noise	•	•	•	•	•	•		•
Land Use and Transport								
Agriculture		•	•	•	•	•	•	•
Residential			•	•	•			
Recreation and tourism		•	•	•				
Industry			•	•				
Minerals and waste				•				
Transport	•		•	•	•	•		

Environmental effects considered in National Grid environmental statements and reports (see Section 2.9)

Index

Printed and bound by CPI Group (UK) Ltd, Croydon, CR0 4YY
16/09/2026
14963379-0003